Revise AS Physics for OCR Sp

David Sang

Heinemann

Contents

Introduction – How to use this revision guide   iii
OCR AS Physics – Assessment   iv

Module A: Forces and motion   1

Block A1: Describing motion
Velocity and displacement   2
Acceleration   4
The equations of motion – part 1   6
The equations of motion – part 2   8
Using vectors   10

Block A2: Explaining motion
Force, mass, acceleration   12
Gravity and motion   14
Force, work and power   16
Turning effect   18

Block A3: Forces in action
Deforming solids   20
Forces on vehicles   22
End-of-module questions   24

Module B: Electrons and photons   27

Block B1: Electric circuits
Electric current   28
Resistance   30
Ohm's law   32
p.d. and e.m.f.   34
Internal resistance and potential dividers   36
Electrical power   38

Block B2: Electromagnetism
Magnetic fields   40
Magnetic forces   42
The ampere and the tesla   44

Block B3: Waves and particles
The electromagnetic spectrum   46
Photons   48
The photoelectric effect   50
Wave–particle duality   52
End-of-module questions   54

Module C: Wave properties   57

Block C1: Rays
Reflection and refraction   58
Total internal reflection   60

Block C2: Wave definitions
Wave representations   62
Wave quantities   64
Wave speed   66

Block C3: Superposition effects
Interference and diffraction   68
Young's slits experiment   70
Superposition and standing waves   72
End-of-module questions   74

Appendices
1: Accuracy and errors   76
2: Data and formulae for question papers   78
3: Useful definitions   80
4: Electrical circuit symbols   82

Answers to quick check questions   83

Answers to end-of-module questions   86

Index   89

Introduction – How to use this revision guide

This revision guide is for the OCR Physics AS course. It is divided into modules to match Specification A. You may be taking a test at the end of each module, or you may take all of the tests at the end of the course. The content is exactly the same.

Each module begins with an **introduction**, which summarises the content. It also reminds you of the topics from your GCSE course which the module draws on.

The content of each module is presented in **blocks**, to help you divide up your study into manageable chunks. Each block is dealt with in several spreads. These do the following:

- they **summarise** the content;
- they indicate **points to note**;
- they include **worked examples** of calculations;
- they include **diagrams** of the sort you might need to reproduce in tests;
- they provide **quick check** questions, to help you test your understanding.

At the end of each module, there are longer **end-of-module questions** similar in style to those you will encounter in tests. **Answers** to all questions are provided at the end of the book.

You need to understand the **scheme of assessment** for your course. This is summarised on page iv overleaf. At the end of the book, you will find a list of the various **formulae** and **definitions** you need to learn, and the others which are provided in tests.

A note about units

In the worked examples, we have included units throughout the calculations. (See for example the worked examples on pages 4 and 5.) This can help to ensure that you end up with the correct units in your final answer. See also the note on checking units on page 7.

OCR AS Physics – Assessment

There are three **units of assessment** (A, B and C) in this AS Physics course. Unit C includes assessment of **experimental skills**.

Unit	Name	Duration of written test	Types of question	Weighting
A	Forces and motion	90 minutes	structured (75 marks); extended (15 marks)	30%
B	Electrons and photons	90 minutes	structured (75 marks); extended (15 marks)	30%
C1	Wave properties	60 minutes	structured (60 marks)	20%
C2	Experimental skills	none	coursework	20%
C3	Experimental skills	90 minutes	practical (60 marks)	20%

> **◖** C2 and C3 are alternatives.

Question types

- **Structured questions** require brief answers to several linked parts of a question.
- **Extended questions** require longer answers to a single question.
- Your answers to the extended questions will be used to assess the quality of your **written communication**.

> **◖** Use the mark allocation and the space available for your answer to guide how much you write.

About the tests

- **Written tests** may be available in January and June.
- **Re-sits** are allowed once only; the better result counts, so you cannot end up with a lower score.
- **Aggregation** means combining the scores for each unit of assessment. You may enter for aggregation at the end of the AS course, or carry your marks forward to the A2 year.
- AS units have half the **weighting** indicated above when they are carried over to the full Advanced GCE award. So AS counts for 50% of A-level.

> **◖** A companion revision guide is available in this series for the A2 part of the course.

Module A: Forces and motion

To help you organise your learning, each module is broken down into blocks. There are three blocks in this module.

- **Block A1** considers how we can describe the motion of an object in terms of displacement, velocity, acceleration and time. This will extend your understanding of motion to show the importance of distinguishing between vector and scalar quantities.

- **Block A2** brings in the idea of forces and how they affect an object's motion. Newton's laws of motion relate force to acceleration. It is important to appreciate that these laws are not self-evident; they go against a lot of what we experience in everyday life.

- **Block A3** applies these ideas in some other situations. Forces can deform materials; understanding how forces affect objects can help to understand safety features of vehicles.

Block A1: Describing motion, pages 2–11

Ideas from GCSE	Content outline of Block A1
• Relationship between speed, distance and time • Graphical representation of speed, distance and time • Acceleration as change in velocity per unit time	• Vector and scalar quantities • Displacement, velocity and acceleration • Graphical representation of motion • Equations of motion • Adding and resolving vectors

Block A2: Explaining motion, pages 12–19

Ideas from GCSE	Content outline of Block A2
• Balanced forces do not alter velocity • Quantitative relation between force, mass and acceleration • Forces on a falling body • Principle of moments • Calculations of work and power	• Force, mass and acceleration • Motion under gravity, air resistance • Force, work and power • Turning effect of a force

Block A3: Forces in action, pages 20–23

Ideas from GCSE	Content outline of Block A3
• Stretching effect of a force • Pressure, force and area • Equal and opposite forces on interacting bodies • Factors affecting stopping distances of vehicles	• Deforming solids, Young modulus • Pressure • Vehicle safety

End-of-module questions, pages 24–26

Velocity and displacement

When an object moves, we may be able to describe its motion using a graph, or an equation. First, we need to define some basic terms.

Speed, distance, time

We can find the **average speed** of a moving object by measuring the distance it travels in an interval of time:

$$\text{average speed} = \frac{\text{distance travelled}}{\text{time taken}}$$

This can only tell us its *average* speed; it may be speeding up or slowing down.

✓ *Quick check 1*

Motion in a straight line

If an object is moving at a steady speed in a straight line, it is in **uniform motion**. Two quantities describe its motion:

- **displacement** – the distance it has travelled in a particular direction;
- **velocity** – its speed in a particular direction.

These are related by the equation:

$$\text{velocity} = \frac{\text{displacement}}{\text{time}} \qquad v = \frac{s}{t}$$

It is important to be able to rearrange the equation for velocity to make time or displacement the subject:

$$\text{displacement} = \text{velocity} \times \text{time} \qquad s = vt$$

$$\text{time} = \frac{\text{displacement}}{\text{velocity}} \qquad t = \frac{s}{v}$$

> ◖ Take care! The symbol s is used for displacement, not speed. Do not confuse it with s for seconds.

SI units In the international system of units (SI units), displacement or distance is measured in **metres** (m), time in **seconds** (s) and velocity in **metres per second** (m s^{-1}). You may come across velocities in a variety of units; keep an eye on the units of displacement and time:

$$\text{m s}^{-1} \qquad \text{mm s}^{-1} \qquad \text{km s}^{-1} \qquad \text{km h}^{-1} \qquad \text{km y}^{-1}$$

> ◖ In the SI system, metres and seconds are *fundamental* or *base* units.

Worked example

A car travels at 25 m s^{-1} for 5 minutes due north along a straight road. What is its displacement after this time?

Step 1 Write down what you know, and what you want to know:

$$\text{velocity } v = 25 \text{ m s}^{-1}, \quad \text{time } t = 5 \text{ min} = 300 \text{ s}, \quad \text{displacement } s = ?$$

Step 2 Choose the form of the equation with displacement as its subject:

$$\text{displacement} = \text{velocity} \times \text{time} \qquad s = vt$$

Step 3 Substitute values and solve:

$$s = 25 \text{ m s}^{-1} \times 300 \text{ s} = 7500 \text{ m} = 7.5 \text{ km}$$

So the car's displacement is 7.5 km due north. Note that, to give a complete answer, we have included the *direction* of the displacement.

✓ *Quick check 2, 3*

Vector and scalar quantities

The definitions of displacement and velocity should remind you of the difference between vector and scalar quantities.

- A **vector quantity** has both magnitude (size) and direction (e.g. displacement, velocity).

- A **scalar quantity** has just magnitude (e.g. distance, speed).

✓ *Quick check 4*

▶▶ *More about the representation of vector quantities on page 10.*

Displacement–time graphs

The shape of an object's **displacement–time graph** shows how its motion is changing. In this example,

1 a straight line sloping up shows that it is going away at a steady speed ('*positive* velocity');

2 a horizontal line shows that it is stationary for a while ('*zero* velocity');

3 a straight line sloping down shows that it is coming back at a steady speed ('*negative* velocity').

In the curved graph the object is speeding up (accelerating); its velocity is positive and increasing.

The **gradient** (slope) of the displacement–time graph is the velocity.
Here, Δ (delta) just means 'change in', so Δs means 'change in displacement'.

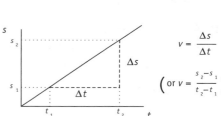

$$v = \frac{\Delta s}{\Delta t}$$

$$\left(\text{or } v = \frac{s_2 - s_1}{t_2 - t_1} \right)$$

❗ Check that the y-axis shows displacement (or distance).

✓ *Quick check 5*

❓ Quick check questions

1 A bus travels along its 20 km route in 40 minutes. Calculate its average speed, and explain why this is only an average.

2 A rocket rises 2000 m vertically upwards in 10 s. What is its average velocity?

3 How long will it take a car travelling at 30 m s^{-1} to travel 1200 m?

4 A spacecraft orbits the Earth at a constant speed of 8 km s^{-1}. Explain whether its velocity is also constant.

5 The table shows how the displacement of a runner along a straight track changes with time. Plot a displacement–time graph. Use it to find the runner's velocity during the first 10 s.

displacement/m	0	45	90	130	172
time/s	0	5	10	15	20

Acceleration

An object is **accelerating** if it is speeding up, **decelerating** if it is slowing down. Its acceleration is a measure of how rapidly its velocity is changing.

Defining acceleration

The **acceleration** of an object is the rate of change of its velocity. If its velocity v changes by an amount Δv in a time interval Δt, its acceleration a is given by:

$$\text{acceleration} = \frac{\text{change in velocity}}{\text{time taken}} \qquad a = \frac{\Delta v}{\Delta t}$$

> Here, Δ (delta) does not represent a quantity. It stands for 'a change in'. So Δv means 'change in velocity'.

We can write this in a different way:

$$\text{acceleration} = \frac{\text{final velocity} - \text{initial velocity}}{\text{time}} \qquad a = \frac{v - u}{t}$$

> We need two symbols for velocity. Remember that u comes before v, so u represents initial velocity.

Units Acceleration is almost always given in m s^{-2} (metres per second squared). It can help to think of an acceleration of, say, 10 m s^{-2} as an increase in velocity of 10 m s^{-1} every second.

Signs An object with a *negative* acceleration (called a deceleration) is slowing down. An object with *zero* acceleration either has uniform velocity (steady speed) or is stationary. *Positive* acceleration means speeding up.

Worked example

A car accelerates from 10 m s^{-1} to 18 m s^{-1} in 4 s. What is its acceleration?

Step 1   Write down what you know, and what you want to know:

$$u = 10 \text{ m s}^{-1}, \quad v = 18 \text{ m s}^{-1}, \quad t = 4 \text{ s}, \quad a = ?$$

Step 2   Write down the equation, substitute and solve:

$$a = \frac{v - u}{t} = \frac{18 \text{ m s}^{-1} - 10 \text{ m s}^{-1}}{4 \text{ s}} = \frac{8 \text{ m s}^{-1}}{4 \text{ s}} = 2 \text{ m s}^{-2}$$

> See the note on units in calculations on page iii.

> ✓ *Quick check 1, 2*

Typical values

It is useful to remember the following values.

- The acceleration in free fall is about 10 m s^{-2}.

- The acceleration of a car or person is usually no more than a few m s^{-2}, as in the worked example.

> The acceleration in free fall is called g – see pages 6 and 14.

Velocity–time graphs

Just as we can draw a displacement–time graph (see page 3), we can draw a **velocity–time graph** to show how an object's *velocity* is changing. In the first graph,

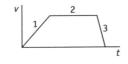

1 a straight line sloping up indicates steadily *increasing* speed (uniform '*positive*' acceleration);

2 a horizontal line shows steady speed ('*zero*' acceleration);

3 a straight line sloping down shows steadily *decreasing* speed ('*negative*' acceleration, or deceleration).

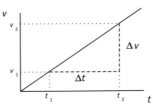

$$a = \frac{\Delta v}{\Delta t}$$

$$\left(\text{or } a = \frac{v_2 - v_1}{t_2 - t_1} \right)$$

The curved graph with decreasing gradient shows decreasing acceleration.

The **gradient** (slope) of the velocity–time graph is the acceleration.

The **area** under the velocity–time graph gives the displacement, because it is the average speed multiplied by the travelling time (see page 6).

❶ Always check the label on the *y*-axis – does the graph show displacement or velocity?

✓ *Quick check 2*

Worked example

Find the displacement after 50 s of the object whose velocity–time graph is shown in the diagram.

Step 1 Divide the area under the graph into a triangle and a rectangle.

Step 2 Calculate the area of each part – see graph.

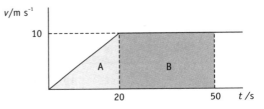

Triangle A: $\frac{1}{2}$ base × height = $\frac{1}{2}$ × 20 s × 10 m s^{-1} = 100 m.

Rectangle B: 30 s × 10 m s^{-1} = 300 m.

Step 3 Add together to give total displacement = 400 m.

✓ *Quick check 3*

❓ Quick check questions

1 An aircraft accelerates from 200 m s^{-1} to 300 m s^{-1} in 25 s. What is its acceleration?

2 A car has an acceleration of 8 m s^{-2}. How long will it take to reach a speed of 24 m s^{-1}, starting from rest?

3 A train travels at a steady speed of 30 m s^{-1} for 100 s. As it approaches a station, it decelerates at a steady rate so that it comes to a halt after 50 s. Draw a velocity–time graph for the train's motion, and use it to calculate

a the train's acceleration as it slows down;

b the total distance travelled during the time described.

❶ You will have to rearrange the equation for acceleration. Alternatively, think of the acceleration as '8 m s^{-1} every second'.

The equations of motion – part 1

The **equations of motion** can be used when an object is accelerating at a steady rate, i.e. its acceleration a is constant. There are *four* equations which need to be learnt. They link five quantities:

u initial velocity v final velocity s displacement

a acceleration t time

The four equations

Here are the four equations, with brief comments. More comments are given with the derivations on pages 8–9.

1	$v = u + at$	This is a rearrangement of $a = \dfrac{v - u}{t}$ (see page 4)
2	$s = \dfrac{v + u}{2} \times t$	This says displacement = average velocity $\times$ time
3	$s = ut + \frac{1}{2}at^2$	With zero acceleration, this becomes displacement = velocity $\times$ time
4	$v^2 = u^2 + 2as$	This is easier to understand in terms of energy and work – see page 16

> For an object falling freely close to the Earth's surface, acceleration $a = g = 9.8$ m s^{-2}, approximately.

❶ These equations only apply to an object moving with uniform (constant) acceleration, and in a straight line.

✓ *Quick check 1*

Worked examples

1 A car travelling at 20 m s^{-1} accelerates at 2 m s^{-2} for 5 s. How far will it travel in this time?

Step 1 Write down what you know, and what you want to know:

$$u = 20 \text{ m s}^{-1}, \quad t = 5 \text{ s}, \quad a = 2 \text{ m s}^{-2}, \quad s = ?$$

Step 2 Choose the appropriate equation linking these quantities:

$$s = ut + \tfrac{1}{2}at^2$$

Step 3 Substitute and solve.

$$s = [20 \text{ m s}^{-1} \times 5 \text{ s}] + [\tfrac{1}{2} \times 2 \text{ m s}^{-2} \times (5 \text{ s})^2]$$

$$= 100 \text{ m} + 25 \text{ m} = 125 \text{ m}$$

Notice that the car's displacement is made up of two parts: 100 m is the distance it would have travelled in 5 s at a steady 20 m s^{-1}; 25 m is the extra distance travelled because it is accelerating.

❶ In rough calculations, you may find it easier to omit the units; however, they provide a check that the quantities are correct. Always include units in examinations.

2 For the car in worked example **1**, use equation 4 to find the car's velocity after it has travelled 125 m (after 5 s). Then use equation 1 to check your answer.

Step 1 Write down what you know, and what you want to know:

$u = 20$ m s^{-1}, $a = 2$ m s^{-2}, $s = 125$ m (from worked example 1), $v = ?$

Step 2 Choose the appropriate equation linking these quantities. The question requires equation 4:

$$v^2 = u^2 + 2as$$

Step 3 Substitute and solve.

$$v^2 = (20 \text{ m s}^{-1})^2 + 2 \times 2 \text{ m s}^{-2} \times 125 \text{ m}$$
$$= 400 \text{ m}^2 \text{ s}^{-2} + 500 \text{ m}^2 \text{ s}^{-2} = 900 \text{ m}^2 \text{ s}^{-2}$$
$$v = \sqrt{900 \text{ m}^2\text{s}^{-2}} = 30 \text{ m s}^{-1}$$

Step 4 Check using equation 1 ($t = 5$ s).

$$v = u + at = 20 \text{ m s}^{-1} + 2 \text{ m s}^{-2} \times 5 \text{ s} = 30 \text{ m s}^{-1}$$

> Note that it is necessary to start a new line when changing from v^2 to v.

> ✓ *Quick check 2–6*

Checking units

In calculations, both numbers and units contribute to the answer. By checking units as you go along, you will have a useful check that you are calculating correctly. For example, in Step 3 of worked example **1** above, we have two terms added together:

$$s = [20 \text{ m s}^{-1} \times 5 \text{ s}] + [\tfrac{1}{2} \times 2 \text{ m s}^{-2} \times (5 \text{ s})^2]$$

First term: units are m s^{-1} × s = m (because s^{-1} and s cancel).

Second term: units are m s^{-2} × s^2 = m (because s^{-2} and s^2 cancel).

❓ Quick check questions

1 An aircraft accelerates at a steady rate from 200 m s^{-1} to 300 m s^{-1} in 80 s. Calculate its acceleration in this time, and its average speed.

2 A stone drops from rest with an acceleration of 9.8 m s^{-2}. How far will it fall in 2.0 s?

> 'From rest' tells you that its initial velocity was zero.

3 A skier moving at a steady speed of 15 m s^{-1} reaches a steeper slope where her acceleration is 1.25 m s^{-2}. How fast will she be travelling after she has moved 160 m from the top of the slope?

4 A train travelling at 10 m s^{-1} accelerates steadily. After 45 s it has reached a speed of 14 m s^{-1}. How far does it travel in this time?

5 At the start of a race, a runner accelerates from rest with a uniform acceleration of 4.5 m s^{-2} for 1.8 s. How fast will she be moving after this time?

6 At a motorway exit, a truck driver brakes from 30 m s^{-1} to 12 m s^{-1} with a deceleration of 2 m s^{-2}. For how long and over what distance is he braking?

> Here a is negative.

The equations of motion – part 2

There are four equations of motion. The first two come from the definitions of velocity and acceleration. The other two can be derived from the first two. This shows that the equations are not independent of one another.

To understand how the equations of motion are related to one another, it is best to start from a graph which represents uniformly accelerated motion. The graph is a straight line. We will also use these ideas (pages 4–5):

- acceleration is the *gradient* of the velocity–time graph;
- displacement is the *area* under the velocity–time graph.

Equation 1: $v = u + at$

The gradient of the graph is the acceleration. Hence we can write

$$a = \frac{v - u}{t}$$

Rearranging gives $at = v - u$, and hence $v = u + at$.

Equation 2: $s = \dfrac{v + u}{2} \times t$

The average velocity is the average of v and u, i.e. $\dfrac{v + u}{2}$.

This is shown by the *dashed* line in the diagram.

The distance travelled is the average velocity × time taken. Hence

$$s = \frac{v + u}{2} \times t$$

This is the rectangular area under the dashed line and is the same as the area under the sloping line.

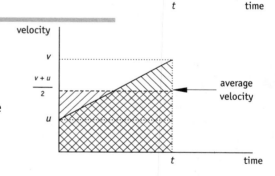

✓ *Quick check 1*

Equation 3: deriving $s = ut + \frac{1}{2}at^2$

We start from equations 1 and 2:

$$v = u + at \qquad \text{and} \qquad s = \frac{v + u}{2} \times t$$

We want a new expression for s. Using equation 1, we substitute for v in equation 2:

$$s = \frac{u + at + u}{2} \times t = \frac{2u + at}{2} \times t$$

Multiplying out gives
$$s = \frac{2u}{2} \times t + \frac{at}{2} \times t$$

and simplifying gives
$$s = ut + \tfrac{1}{2}at^2.$$

Alternative derivation

This is the same graph as before. The area under the graph is the displacement, and is made up of two parts: a rectangle and a triangle.

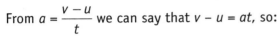

The rectangle represents the displacement of an object moving with velocity u for time t:

Area of rectangle = width × height = ut

The triangle represents the extra distance travelled because the object is speeding up:

Area of triangle = $\frac{1}{2}$ base × height = $\frac{1}{2}t \times (v - u)$

From $a = \dfrac{v-u}{t}$ we can say that $v - u = at$, so:

Area of triangle = $\frac{1}{2}t \times (v - u) = \frac{1}{2}t \times at = \frac{1}{2}at^2$

Adding the two areas gives the total displacement:

$$s = ut + \tfrac{1}{2}at^2$$

✓ *Quick check 2*

Equation 4: deriving $v^2 = u^2 + 2as$

Starting from $v = u + at$, we have $v - u = at$.

Starting from $s = \dfrac{v+u}{2} \times t$, we have $v + u = \dfrac{2s}{t}$.

Multiplying these two equations together gives

$$(v + u) \times (v - u) = \frac{2s}{t} \times at = 2as$$

The left-hand side is the difference of two squares, i.e. $v^2 - u^2$. So $v^2 - u^2 = 2as$ and rearranging gives $v^2 = u^2 + 2as$.

✓ *Quick check 3*

? *Quick check questions*

1 Car A is travelling at a steady speed of 15 m s^{-1}. It overtakes car B, which is travelling at 10 m s^{-1}. Car B's driver immediately starts to accelerate uniformly. Car B catches up with car A after 20 s. *On the same axes*, draw velocity–time graphs to represent the motion of the two cars. What is car B's speed when it catches up with car A? Use your graph to show that the cars travel 300 m before B catches up with A.

2 A car, initially travelling at 7 m s^{-1}, accelerates steadily for 10 s until it reaches a speed of 12 m s^{-1}. Draw a velocity–time graph to represent this motion. Use the graph to deduce the car's acceleration, and how far it travels.

➤ Check your answer using $s = \dfrac{u+v}{2} \times t$.

3 A spacecraft travelling in a straight line accelerates from 8 km s^{-1} to 12 km s^{-1} with an acceleration of 1.6 km s^{-2}. How far does it travel whilst accelerating?

➤ You will have to rearrange $v^2 = u^2 + 2as$.

Using vectors

Vector quantities, such as displacement, velocity and acceleration, have both *magnitude* and *direction*. They cannot simply be represented by a numerical value. Instead, to include their direction, they can be represented by drawing **vector diagrams**. In a vector diagram:

- the *length* of a line represents the magnitude of the vector quantity;
- the *direction* of a line represents the direction of the vector quantity.

Vector diagrams are a type of scale drawing.

Reminder of Pythagoras' theorem: for a right-angled triangle, $a^2 + b^2 = c^2$.

Adding two vectors

Draw a **vector triangle** as follows.

- Choose a suitable scale. Draw a line (AB) to represent the first vector quantity. Add an arrow to the line.
- From the end of the first vector, draw a line (BC) to represent the second. Add an arrow.
- Return to the *start* of the first vector. Draw a line from this point to the *end* of the second vector. This line (AC) represents the **resultant** of the two. Add *two* arrowheads to show that this is the resultant of the other two vectors. This process is called **adding two vectors**.

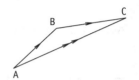

You can think of this as 'Going from A to B to C is the same as going directly from A to C'.

Worked example

When the two vectors are perpendicular, the triangle is right-angled and the resultant is the hypotenuse. Use Pythagoras' theorem to work out the resultant.

A ship is sailing due north at 8 m s^{-1}. A passenger walks across the deck at 4 m s^{-1}, in an easterly direction. What is her resultant velocity?

The diagram shows how the two velocities are added.

Step 1 Draw a vector to represent the velocity of the ship.

Step 2 Draw a vector to represent the passenger's velocity across the deck.

Step 3 Draw the resultant vector.

Step 4 Measure or calculate the resultant. In this case, we have a right-angled triangle, so we can calculate the resultant using Pythagoras' theorem.

$$v^2 = (8 \text{ m s}^{-1})^2 + (4 \text{ m s}^{-1})^2 = (64 + 16) \text{ m}^2 \text{ s}^{-2} = 80 \text{ m}^2 \text{ s}^{-2}$$

$$v = \sqrt{80 \text{ m}^2 \text{s}^{-2}} = 8.9 \text{ m s}^{-1}$$

We must also state the *direction* of the passenger's resultant velocity. We need to find the angle θ from the diagram.

$$\tan \theta = \frac{\text{opp}}{\text{adj}} = \frac{4}{8} = 0.5$$

$$\theta = \tan^{-1} 0.5 = 26.6°$$

The passenger's resultant velocity is thus 8.9 m s^{-1} at 26.6° east of north.

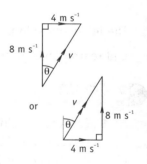

It is always advisable to check a calculation like this by drawing a scale diagram and measuring both the length of the resultant, and the angle.

✓ *Quick check 1*

Resolving a vector

Sometimes it is useful to **resolve** (break down) a vector quantity into two **components** at 90° to one another.

resolving

Imagine turning the vector V round to point in the direction of interest. If you turn the vector through an angle θ, the component of V in this direction is $V \cos \theta$.

A vector may be replaced by two perpendicular components whose values are $V \cos \theta$ and $V \sin \theta$. Notice that if these two components are *added* (see page 10), the resultant is the original vector.

two perpendicular components

The *perpendicular components* of a vector are *independent* of one another. Changing one component has no effect on the other.

adding perpendicular components

✓ *Quick check 2*

Worked example

A car is travelling at 20 m s^{-1} at 30° W of N (see diagram). Calculate the components of its velocity due N and due W.

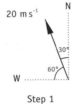 

Step 1   Step 2   Step 3

Step 1   Draw a diagram; mark the relevant angles.

Step 2   Calculate the component due N. (The angle here is 30°.)

       Component due N = 20 m s^{-1} × cos 30° = 17.3 m s^{-1}

Step 3   Calculate the component due W. (The angle here is 60°.)

       Component due W = 20 m s^{-1} × cos 60° = 10.0 m s^{-1}

● Note that we could have calculated this last result using sin 30° rather than cos 60°.

✓ *Quick check 3*

? *Quick check questions*

   1   A whale swims 1000 km due S and then 400 km due E. How far is it from its starting point, and in what direction?

   2   The Earth's gravity makes objects fall with an acceleration of g = 9.8 m s^{-2}. What will be the acceleration of an object down a smooth slope, inclined at 45° to the horizontal?

   3   A boy is speeding at 8 m s^{-1} down a water slide which is inclined at 35° to the horizontal. Calculate the horizontal and vertical components of his velocity.

● Calculate the component of g at 45° to the horizontal.

Force, mass, acceleration

So far, we have only *described* motion. Now we can go on to use the idea of forces to *explain* why motion changes. Newton's laws of motion tell us about how forces change motion:

- If the forces on an object are *balanced*, it will remain at rest or continue to move with constant velocity (i.e. at a steady speed in a straight line).

- If the forces on an object are *unbalanced*, its motion will change; it will accelerate in the direction of the unbalanced force.

Force and acceleration

Force, mass and acceleration are related by the equation:

> **force = mass × acceleration $F = ma$**

Here, F is the **unbalanced force** (the resultant force) in newtons (N) acting on an object of mass m in kilograms (kg). It gives the object an acceleration a in m s^{-2} which is *in the direction of the force*. See 'Units: the newton' below.

The equation $F = ma$ is often used as a shorthand way of remembering Newton's second law of motion, but don't forget that *direction* is important too.

To sum up, an object acted on by an unbalanced force will accelerate. Its acceleration is proportional to the force and is in the direction of the force.

✓ *Quick check 1*

Worked example

A car of mass 1000 kg is acted on by two forces: a forward force of 500 N provided by its engine, and a retarding (backward) force of 200 N caused by air resistance. What is its acceleration?

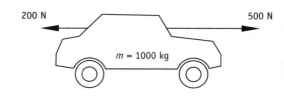

Step 1   Draw a diagram to show the forces acting on the car. (It can help to draw a longer arrow for the larger force.)

Step 2   Calculate the unbalanced force and note its direction.

> $F = 500$ N – 200 N = 300 N **forwards**

Step 3   Calculate the acceleration by rearranging $F = ma$.

$$a = \frac{F}{m} = \frac{300 \text{ N}}{1000 \text{ kg}} = \frac{300 \text{ kg m s}^{-2}}{1000 \text{ kg}} = 0.3 \text{ m s}^{-2}$$

So the car's acceleration is 0.3 m s^{-2} forwards.

❶ Don't forget to state the direction.

✓ *Quick check 2, 3*

Units: the newton

The unit of force is the **newton**. The equation $F = ma$ defines the newton:

$$1 \text{ N} = 1 \text{ kg} \times 1 \text{ m s}^{-2}$$

A newton is the force that will give a mass of 1 kg an acceleration of 1 m s^{-2}. Equally, it will give a mass of 0.5 kg an acceleration of 2 m s^{-2}, and so on.

The kilogram, metre and second are **fundamental units** in the SI system. The newton is a **derived unit**. Most units we use are derived units; it is important to be able to trace them back to the fundamental units.

▶▶ *More about the relationships between units in the SI system on pages 2 and 80–81.*

The meaning of mass

The equation $F = ma$ also tells us what we mean by '**mass**'. Imagine the same unbalanced force acting on two objects, one of large mass, the other of small mass. The object with larger mass will accelerate less than the object with smaller mass.

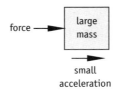

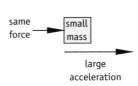

Mass is the property of an object which resists change in motion.

In diagrams, it often helps to label an object with its mass, but don't put an arrow. Mass does not have direction – it is a *scalar* quantity.

❶ Don't confuse mass with weight – see page 14.

✓ *Quick check 4, 5*

? Quick check questions

1 Is force a vector or a scalar quantity?

2 What force is needed to give a ball of mass 2.4 kg an acceleration of 15 m s^{-2}?

3 A parachutist of mass 80 kg and weight 800 N is acted on by an upward drag force of 960 N. What is his acceleration?

4 Is mass a vector or a scalar quantity?

5 When force F acts on object A, it is given an acceleration of 13 m s^{-2}. When the same force F acts on object B, its acceleration is 14 m s^{-2}. Which has the greater mass, A or B?

❶ Don't forget to state its direction.

Gravity and motion

The Earth's gravitational pull on us causes us to have *weight*. This is a force which acts on us all the time, so that we hardly notice its existence. Because our weight is proportional to our *mass*, we tend to get the two ideas confused.

Gravity and acceleration

If an object falls freely, it accelerates downwards. For objects near the surface of the Earth, the acceleration caused by gravity has the *approximate* value $g = 9.8$ m s^{-2}. This decreases the further you go from the Earth's centre, so the value of g varies over the Earth's surface. It is greater nearer the poles where the Earth is slightly flattened, and less nearer the equator, as well as at higher altitudes.

In each succeeding second, the object falls further. It is accelerating downwards.

$t = 0$

$t = 1$ s

$t = 2$ s

$t = 3$ s

✓ *Quick check 1*

Gravity and weight

The fact that a falling object accelerates shows that there must be an unbalanced force acting on it – its **weight**. An object's weight depends on two factors:

- its mass m (the greater the mass, the greater its weight);
- the gravitational field strength g.

weight = mass × gravitational field strength $\quad$ $W = mg$

The **gravitational field strength** g tells you how many newtons of force pull on each kilogram of mass. It has the approximate value $g = 9.8$ N kg^{-1} (newtons per kilogram) near the Earth's surface.

The acceleration caused by gravity $g = 9.8$ m s^{-2} and the gravitational field strength $g = 9.8$ N kg^{-1} are two different ways of saying the same thing. 1 N kg^{-1} is the same as 1 m s^{-2}. On the surface of the Moon, gravity is much weaker. The field strength is 1.6 N kg^{-1}, so falling objects have an acceleration of 1.6 m s^{-2}.

✓ *Quick check 2*

Weight and mass

Weight is a force, measured in newtons. It is represented by an arrow. The weight of an object depends on where it is, because gravitational field strength varies from place to place.

Mass is a property of an object. It tells us how much matter it is made of. It is a measure of resistance to change in motion. It does not vary from place to place.

✓ *Quick check 3*

Falling through air

An object falling through air experiences air resistance, known as **drag**. This is a resistive force, opposing motion, and is always in the opposite direction to velocity.

Air resistance increases as an object moves faster. Eventually, air resistance equals the object's weight, and the forces are balanced. The object cannot go any faster; it has reached **terminal velocity**.

This explains why cars, for example, have a top speed. The forward force of the engine is matched by the backward force of air resistance. To go faster, the car would need an engine which provided a greater force; alternatively, it could be redesigned to reduce air resistance.

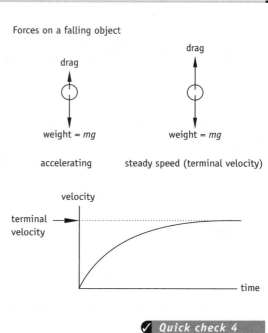

Forces on a falling object

accelerating   steady speed (terminal velocity)

✓ *Quick check 4*

Projectile motion

An object thrown in any direction except vertical follows a curved (parabolic) path through the air. Gravity acts on it vertically, and it has no horizontal acceleration.

- Horizontal motion: constant velocity.
- Vertical motion: uniform acceleration g.

Resolve the initial velocity into horizontal and vertical components and treat them separately.

? Quick check questions

1 How far will an object fall, starting from rest, in 3 s? Its acceleration is 9.8 m s^{-2}.

2 The gravitational field strength on the surface of Mars is 3.8 N kg^{-1}. What will an object of mass 60 kg weigh there? How far will it fall, starting from rest, in 3 s?

3 Is weight a vector or a scalar quantity? Is mass a vector or a scalar quantity?

4 A skydiver is falling at a steady speed of 50 m s^{-1}. He opens his parachute so that the force of air resistance increases. Describe and explain how his motion will change after this.

Force, work and power

We use forces to do things – to start things moving, to make them accelerate or decelerate, to change their shape. When a force changes the *energy* of something in this way, we say that it does **work**. Energy has been *transferred*.

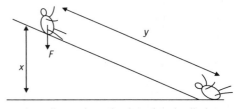
gravity pulls you straight down: $W = Fx$

work done = energy transferred

Doing work

When a force F pushes an object for a distance s, the work W done by the force is

work done = force $\times$ displacement *in the direction of the force* $W = Fs$

Although both force and displacement in a particular direction are vector quantities, their product (work done) is a scalar. Energy and work are both scalar quantities.

C Take care! The displacement s must be measured along the direction of the force.

✓ *Quick check 1*

Defining the joule

Since work done tells us how much energy is transferred by a force, both work and energy are measured in the same units, called **joules** (J). The equation $W = Fs$ relates joules to newtons and metres.

1 joule = 1 newton $\times$ 1 metre   1 J = 1 N m

1 joule is the energy transferred when a force of 1 newton moves through 1 metre. Equally, it is the energy transferred when a force of 0.5 N moves through 2 m, and so on. So, in fundamental units, $1\ \text{J} = 1\ \text{N m} = 1\ \text{kg m s}^{-2}\ \text{m} = 1\ \text{kg m}^2\ \text{s}^{-2}$.

gravity pulls you down the slope, but the displacement in the direction of the force is x, not y: $W = Fx$

✓ *Quick check 2*

Kinetic energy

A force F applied to an object of mass m initially at rest ($u = 0$) gives it an acceleration a. After it has moved a distance s, the fourth equation of motion (page 6) tells us that its velocity will be given by

$$v^2 = 2as$$

Multiplying both sides by $\frac{1}{2} m$ gives

$$\tfrac{1}{2} mv^2 = mas = Fs$$

since $F = ma$. Now Fs is the work done by F acting over the distance s, that is, the energy given to the moving object. This is called the object's **kinetic energy** (KE) and is measured in joules (J).

KE $= \frac{1}{2} mv^2$

Notice that this has the correct units of $\text{kg m}^2\ \text{s}^{-2}$. If $u \neq 0$, then

final KE = initial KE + work done by force

Gravitational potential energy

When an object of weight mg is raised through a height Δh, work is done against gravity. The energy given to the object as a result is called its **gravitational potential energy** (GPE). This is increased by an amount ΔE_p given by

> **gain in GPE = weight × gain in height** $\Delta E_p = mg\Delta h$

The symbol Δ means 'a change in ...', so Δh means 'a change in height h'. It is important to think of Δh as a single mathematical symbol, not two quantities multiplied together.

Worked example

A stone falls from a height of 5 m. How fast is it moving when it reaches the ground?

Step 1   The decrease in the stone's GPE as it falls is equal to its gain in KE.

$$mg\Delta h = \tfrac{1}{2} mv^2$$

Step 2   Cancel m from both sides: $g\Delta h = \tfrac{1}{2} v^2$

Step 3   Substitute values and solve for v: $9.8 \text{ m s}^{-2} \times 5 \text{ m} = 0.5v^2$

$$v^2 = 98 \text{ m}^2 \text{ s}^{-2}$$
$$v = 9.9 \text{ m s}^{-1}$$

Note that the fact that m cancels out means that we would get the same answer for any value of m, i.e. all stones of whatever mass would fall at the same rate (neglecting air resistance).

✓ *Quick check 3*

Power

Power P is the *rate of doing work*, or the *rate of transferring energy*:

> $$\text{power} = \frac{\text{work done}}{\text{time taken}} \qquad P = \frac{W}{t}$$

Power is measured in **watts**, W.

> **1 watt = 1 joule per second $1 \text{ W} = 1 \text{ J s}^{-1}$**
> **1 kilowatt = 1 kW = 1000 J s^{-1}**

✓ *Quick check 4, 5*

❓ Quick check questions

1 A car's engine provides a forward force of 500 N. It is opposed by a resistive force of 200 N. The car accelerates forwards for 20 m. How much work is done by each force? By how much does the car's energy increase?

2 Show that $1 \text{ J} = 1 \text{ kg m}^2 \text{ s}^{-2}$.

3 A car of mass 1000 kg is travelling at 20 m s^{-1}. What is its kinetic energy? It climbs a hill 200 m high. By how much does its gravitational potential energy increase?

4 The engines of a light aircraft provide a power of 50 kW. How much energy do they transfer in 1 minute?

5 Which of the following are vector quantities: force, kinetic energy, gravitational potential energy, work done, power?

Turning effect

Forces can have many different effects on the objects they act on. The **moment** of a force tells us about its *turning effect*.

Moment of a force

The moment of a force about a point (the *pivot*) is defined as:

> **moment = *magnitude* of the force × perpendicular *distance* of its line of action from the pivot**

It is important to be able to determine the distance between the point and the line of action (direction) of the force. The worked example illustrates another method.

Units   Moment is measured in **newton-metres** (N m).

Note that 1 N m is not the same as 1 J. In calculating work done (in J), force and distance are in the *same* direction. In calculating moment (in N m), force and distance are *perpendicular*.

F and *d* perpendicular:
moment = $F \times d$

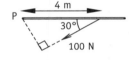

Draw a line from point P at 90° to line of force:
moment = $F \times d$

Worked example

A force of 100 N acts at an angle of 30° to a beam, and at a distance $x = 4.0$ m from one end. What is the moment of the force about this end?

Method 1
Step 1   Draw the line of action of the force. Then draw a perpendicular from P to the line of action.

Step 2   Calculate the length *d* of this line:

$$d = 4.0 \text{ m} \times \sin 30° = 2.0 \text{ m}$$

Step 3   Multiply by the force to find the moment:

$$\text{moment} = F \times d = 100 \text{ N} \times 2.0 \text{ m} = 200 \text{ N m}$$

Method 2
Step 1   Calculate the component of *F* perpendicular to *x*:

$$\text{component of } F = 100 \text{ N} \times \cos 60° = 50 \text{ N}$$

Step 2   Calculate the moment of this component about P:

$$\text{moment} = 50 \text{ N} \times 4.0 \text{ m} = 200 \text{ N m}$$

Of course, both methods give the same answer.

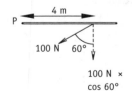

✓ *Quick check 1*

Torque of a couple

A **couple** is a pair of forces. They are equal in magnitude, and act in opposite directions, but they do not lie in the same line. Because of their equal and opposite sizes, they do not make the object accelerate away. However, because they do not line up, they tend to make the object *rotate*. The moment of a couple is known as its **torque** (units: N m).

> **torque of a couple = magnitude of one force × perpendicular distance between them**

✓ *Quick check 2*

Equilibrium

If an object is **in equilibrium**, it will not accelerate, and it will not start to rotate. For this to happen:

● there must be no resultant (unbalanced) force acting on it;

● there must be no resultant torque acting on it.

✓ *Quick check 3*

Centre of gravity

An object may have a complicated shape; gravity acts on all parts of it. Every object has a single point, called its **centre of gravity**, on each side of which the moments of all the separate parts of the object are balanced.

We can represent the weight of the whole object by a downward arrow acting at its centre of gravity. This greatly simplifies problem solving.

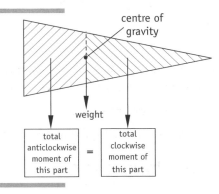

Density

The **density** of a substance is its mass per unit volume. Symbol ρ (Greek letter rho).

$$\rho = \frac{m}{V}$$

✓ *Quick check 4*

? *Quick check questions*

1 Calculate the moment about P of each of the two forces shown in the upper diagram.

2 Which two of the forces shown in the lower diagram constitute a couple? What is their torque?

3 Copy the diagram for question **2**. Add another force, acting at point X, which will leave the object in equilibrium. Show that there is no resultant force or torque acting on it.

4 What is the density of a substance if 24 kg occupies 0.04 m³? What volume would be occupied by 1200 kg of the substance?

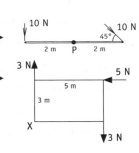

Deforming solids

It takes a pair of forces to *stretch* a solid object; such forces are called **tensile**. Forces which squash or *compress* an object are called **compressive**. Tensile forces can stretch and break an object. It is easiest to start by describing how a spring stretches.

Hooke's law

The greater the **load** (the force stretching the spring), the greater its **extension** (increase in length). Eventually the load is so great that the spring becomes permanently stretched. The graph shows two things:

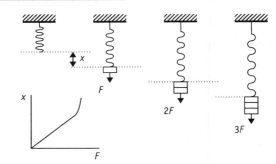

- At first the graph is a straight line, so load F is proportional to extension x: $F \propto x$.

- Beyond the **elastic limit**, the spring does not return to its original length when the load is removed.

The proportionality can be turned into an equation; the constant of proportionality is called the **spring constant** k.

$$F = kx$$

k is measured in N m^{-1}. It is sometimes known as the *stiffness* of the spring. k tells you how many newtons are needed to stretch the spring by 1 metre.

The area under the graph is $\frac{1}{2} \times$ force $\times$ distance; this tells you the work done in stretching the spring, which is known as the **strain energy**.

> ❶ You may find 'stiffness' easier to remember than 'spring constant'.

> ✓ *Quick check 1*

Stretching a wire

Measure the original length of the test wire using a metre rule; measure its diameter using a micrometer, and calculate its cross-sectional area.

Gradually increase the load on the test wire. For each load, note the reading from the vernier scale.

Calculate values of stress and strain from values of load and extension, and plot a stress–strain graph.

The extension x of the wire depends on four factors:

- its original length l,
- the load F stretching it,
- the cross-sectional area A,
- the stiffness of the material of which it is made.

These are combined as follows:

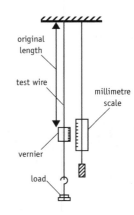

> strain = extension/original length: strain = x/l
> stress = load/area: stress = F/A

The greater the stress, the greater the strain which results. For many materials:

$$\frac{\text{stress}}{\text{strain}} = \text{constant}$$

and this constant is known as the **Young modulus** Y of the material. Y is measured in N m^{-2} or **pascals** (Pa). Its value is typically expressed in MPa or GPa (millions or billions of pascals); 1 MPa = 1 megapascal = 10^6 Pa; 1 GPa = 1 gigapascal = 10^9 Pa.

$$\text{Young modulus} = \frac{\text{stress}}{\text{strain}}$$

✓ *Quick check 2*

Stress–strain graphs

For low stresses, below the elastic limit (e.l.), the material will return to its original length when the load is removed. This is **elastic deformation**. For higher stresses, the material becomes permanently deformed. This is **plastic deformation**.

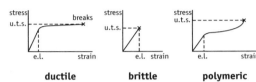

The Young modulus is the gradient of the *initial* (straight line) part of the graph. Once the stress reaches its highest value (the **ultimate tensile stress**, u.t.s.) the material will break.

Materials may be classified as follows:

- A **ductile** material stretches a lot beyond the elastic limit, e.g. copper.
- A **brittle** material snaps when it reaches the elastic limit, e.g. glass.
- A **polymeric** material does not show linear behaviour, e.g. polyethene.

✓ *Quick check 3*

Pressure

Pressure is defined as force per unit area: $P = \dfrac{F}{A}$

It is measured in N m^{-2}, also known as **pascals** (Pa).

The greater the force and the smaller the area it acts on, the greater the pressure.

C You can think of stress as the same as pressure; each is a force divided by an area.

✓ *Quick check 4*

❓ Quick check questions

1 A spring is 1.2 m long when unstretched, and 1.4 m long when a load of 50 N is applied. Its elastic limit is 80 N. Calculate the extension produced by the 50 N load, and the spring constant.

2 A stress of 20 MPa is applied to a wire of Young modulus 10 GPa. What strain is produced? If the wire's initial length was 0.8 m, what extension does the stress produce?

3 The table shows stress–strain data for an iron wire. Use the data to plot a stress–strain graph; deduce the Young modulus for iron; and state whether the wire shows plastic or brittle behaviour.

stress/MPa	0	50	100	150	200	250
strain/%	0	0.025	0.050	0.075	0.100	wire breaks

4 The pressure of the atmosphere is 10^5 Pa. What force does it exert on a child of surface area 1.5 m^2?

Forces on vehicles

We can use the ideas of force, work and power to understand how vehicles change speed. The same ideas can be used to understand factors affecting car safety.

Motive force

The engine of a car provides the force that makes it accelerate. This is the **motive force**.

frictional force of tyre on road   frictional force of road on tyre

The engine drives the wheels round. The bottom of the tyre (where it is in contact with the road surface) is trying to move *backwards*. It pushes backwards on the road; provided the road is not slippery, there is a frictional force *forwards*, exerted by the road on the tyre. It is this force which propels the car forwards.

These two forces:

- the backward push of the tyre on the road
- the forward push of the road on the tyre

are an example of a pair of forces subject to Newton's third law of motion: *action* and *reaction* are equal and opposite. They are both the same type of force (frictional); they are equal in magnitude but opposite in direction; they act on different objects.

Car tyres must have good tread so that they make good contact with the road surface. This ensures that the road can provide a large motive force to push the car and make it accelerate. Similarly, the road surface must provide good grip so that the tyres do not slip and cause the car to skid.

✓ *Quick check 1*

Motive power

The **motive power** of the engine is the rate at which it transfers energy to the car. To calculate the motive power, we use

> **motive power = motive force × speed**

This is the same as **force** $\times \dfrac{\textbf{distance}}{\textbf{time}}$, or $\dfrac{\textbf{work done}}{\textbf{time}}$ (see page 17).

It is measured in **watts** (W).

> $$1 \text{ W} = 1 \text{ N} \times \text{m s}^{-1} = 1 \text{ J s}^{-1}$$

The motive power of a car is likely to be tens of kilowatts (kW); the engine itself generates much more power than this, but a large fraction is wasted as heat.

✓ *Quick check 2*

Braking force

The frictional drag of the air tends to slow a car down. To stop more quickly, the driver applies the brakes. These produce a frictional force on the wheels or axles to slow their rotation. This is the **braking force**.

speeding up　　steady speed　　braking

drag　motive force　drag　motive force　braking force

drag

Stopping distances and safety features

If a car has to stop suddenly, it takes a fraction of a second for the driver to react – this is the *thinking time*, in which the car travels a short distance called the **thinking distance**. Once the brakes are applied, the car travels a further distance called the **braking distance** before coming to a halt.

stopping distance = thinking distance + braking distance

If the car's tyres are bald, and if the road surface is smooth or wet, friction will be reduced and the braking distance will be greater.

When a car comes rapidly to a halt, for example in a crash, its kinetic energy decreases to zero. Work is done in crushing the **crumple zone**, and this absorbs the energy. **Seat belts** and **air bags** bring the driver gradually to rest, avoiding a sudden sharp impact with the windscreen or dashboard.

✓ *Quick check 3*

? *Quick check questions*

1 A car travels 100 m along a road. It exerts a frictional force of 500 N on the road. Calculate the work done by this force on the road, and the work done by the force of the road on the car.

2 A car of mass 1000 kg is travelling at 20 m s^{-1}. It accelerates at 2 m s^{-2}. Calculate the motive force which is needed to give it this acceleration, and the motive power of the engine.

3 A driver travelling at 15 m s^{-1} sees an obstruction ahead. She reacts quickly and applies the brakes after 0.6 s. The car decelerates to rest at 3 m s^{-2}. Calculate the thinking distance, braking distance and stopping distance.

❶ Use $v^2 = u^2 + 2as$ to calculate the stopping distance.

Module A: end-of-module questions

1 a What is the essential difference between speed and velocity?

b Which of the following quantities are vectors, and which are scalars?

force, velocity, distance, acceleration, kinetic energy, power

2 a Write down a word equation which defines acceleration.

b The graph shows how the velocity of a car varied during the first part of a journey. Calculate the car's acceleration during the parts of the journey represented by AB and BC.

c Copy the graph and add to it to indicate how you would expect the car's velocity to change if it was involved in a sudden collision with a brick wall.

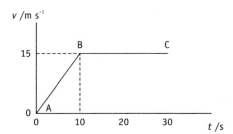

3 The diagram shows a heavy load being dragged by two tractors. The tension in each cable is 20 kN.

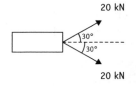

a Draw a vector triangle and use it to find the resultant of these two forces.

b The load moves at a steady speed along the ground. What can you say about the resistive force acting on it?

4 When the space shuttle comes in to land, it deploys a parachute to slow it down. The graph shows how the horizontal force F on the shuttle varies with time t.

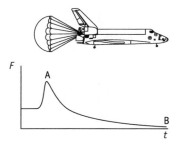

a The parachute is deployed at point A on the graph. Explain why the force on the shuttle decreases in the region AB.

b Copy the graph. On the same axes, add a further line to show how the shuttle's velocity changes.

5 a 'The gravitational field strength g at the Earth's surface is 9.8 N kg^{-1}.' Explain what is meant by this statement.

b An astronaut measures his weight on Earth as 882 N. He travels to the surface of Mars, where the gravitational field strength is 3.8 N kg^{-1}. What will be the values of his mass and weight on Mars?

c Show that 9.8 N kg^{-1} is the same as 9.8 m s^{-2}.

6 a With the aid of a diagram, explain what is meant by the moment of a force.

The diagram shows a diving board which projects horizontally from point X. Its weight (2000 N) acts half-way along its length, as shown. It is supported by a cable attached 1 m from the far end, and which makes an angle of 50° with the vertical.

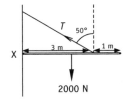

b Calculate the moment of the board's weight about point X.

c Show that the tension *T* in the cable must be 2074 N if its moment is to counteract the moment of the board's weight.

7 An experiment is conducted on a rubber cord of unstretched length 2.0 m. The cord has a rectangular cross-section of dimensions 0.5 mm by 5 mm. The cord is hung vertically and a weight of 100 N attached to its lower end. Its length increases to 2.4 m. Calculate:

a the stress in the cord

b the strain

c the Young modulus of the rubber.

8 A spring is stretched by hanging increasing weights on its end, and its length is measured for each value of the load. The table shows the results.

load/N	0	10	20	30	40
length/m	0.060	0.062	0.064	0.066	0.068

a Plot a suitable graph and use it to determine the spring constant of the spring.

b The final load of 40 N does not exceed the spring's elastic limit. Explain what is meant by *elastic limit*.

c Use the graph to calculate the strain energy stored when the spring is stretched by the 40 N load.

9 Police investigating a road accident discover that one of the cars has bald tyres; most of the tread has been worn away. They suggest to the driver that he should have noticed problems when he attempted to make the car accelerate.

a Draw a diagram to show the horizontal forces which act between the road surface and the tyre of a car. Use your diagram to explain the origin of the motive force which causes a car to accelerate.

b Explain why a bald tyre will result in a reduced motive force.

c The driver claims that he was travelling at 20 m s^{-1} when he braked. The police say that he should have been able to stop in a distance of 40 m if his tyres had been roadworthy. What would the driver's deceleration have been if he had stopped in this distance?

d Police tests show that the driver's bald tyres could not have provided a deceleration greater than 2 m s^{-2}. What braking distance would be required if the car was travelling at 20 m s^{-1}?

10 An aircraft of mass 100×10^3 kg is moving at 150 m s^{-1}. It reduces its height above the ground from 5000 m to 3000 m and increases its speed to 200 m s^{-1}. Calculate the change in its gravitational potential energy, and in its kinetic energy. (Gravitational field strength = 9.8 N kg^{-1}.)

11 A boy stands on the edge of a cliff. He throws a stone vertically upwards so that it leaves his hand, 2.0 m above the clifftop, with an initial velocity of 8.0 m s^{-1}. It rises upwards and then falls vertically downwards to the foot of the cliff. (Gravitational field strength = 9.8 N kg^{-1}.)

(You may assume that air resistance is negligible throughout the stone's motion.)

   a Calculate the greatest height to which the stone will rise (above the point where it leaves the boy's hand).

   b If the cliff is 20 m high, calculate the stone's velocity when it reaches the foot of the cliff.

   c What is the stone's average velocity during its motion?

   d How much time elapses between the moment when the stone leaves the boy's hand and when it reaches the foot of the cliff?

12 a What is the SI unit of force?

   b Write an equation to show how this unit is related to the kilogram, metre and second (three of the base units of the SI system).

13 a Sketch a force–extension graph for a ductile material.

   b Indicate the regions in which you would expect its behaviour to be elastic, and where it will be plastic.

   c State the essential difference between elastic and plastic deformation of a solid material.

Module B: Electrons and photons

There are three blocks in this module.

- **Block B1** is the longest. It concerns electricity, and it will help you to solve problems involving current, voltage and resistance, as well as energy and power in electric circuits.

- **Block B2** will extend your understanding of magnetism and its relation to electricity. You will learn how to describe the strength of a magnetic field in terms of flux density.

- **Block B3** introduces some ideas from modern physics about the mysterious, fundamental nature of matter and radiation. We are used to thinking of matter as made up of particles, and radiation as waves. Here you will learn that things are not so simple. Sometimes matter behaves as a wave; and radiation may behave as particles.

Block B1: Electric circuits, pages 28–39

Ideas from GCSE	Content outline of Block B1
• Relationship between current, voltage and resistance • Measuring current and voltage • Nature of electric current • Energy and power in electric circuits	• Electric current and potential difference • Resistances in series and parallel • Ohm's law • Resistance and resistivity • Electromotive force and internal resistance • Kirchhoff's laws • Energy transfers in electric circuits

Block B2: Electromagnetism, pages 40–45

Ideas from GCSE	Content outline of Block B2
• Magnetic fields produced by magnets and electric currents • Magnetic forces of attraction and repulsion • Electric motors	• Magnetic fields of currents in wires and coils • Force on a current-carrying conductor in a magnetic field • Magnetic flux and flux density • Definitions of the ampere and tesla

Block B3: Waves and particles, pages 46–53

Ideas from GCSE	Content outline of Block B3
• Reflection, refraction and diffraction of waves • Electromagnetic spectrum • Speed, frequency and wavelength of a wave	• Electromagnetic spectrum • Line spectra and energy levels in atoms • Photons and photon energy • Photoelectric effect • Electron diffraction • Wave–particle duality

Studying Chemistry? You may already know about emission and absorption spectra, and how each element can be identified by the wavelengths of light in its spectrum.

End-of-module questions, pages 54–56

Electric current

A battery pushes charge around a circuit. A current is a flow of *positive* charge. Because of this, current must flow from positive to negative. This is known as **conventional current**.

The charged particles in metals are **electrons**. Electrons are negatively charged, so they flow from negative to positive. We still say that the current flows from positive to negative.

▶▶ *More about electrons in metals on pages 50 and 52.*

Conservation of current

Current is the flow of charge. Charge cannot disappear or get used up. For this reason, we say that **current is conserved**.

- At point X, current splits up: $I = I_1 + I_2$
- At point Y, currents recombine: $I_1 + I_2 = I$

These equations describe the conservation of current. They are an example of **Kirchhoff's first law**. A more formal statement is:

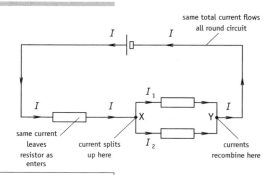

> **The sum of the currents entering a point is equal to the sum of the currents leaving the point.**
> $\Sigma I_{in} = \Sigma I_{out}$ where Σ (sigma) means 'the sum of'.

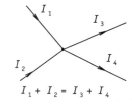

$$I_1 + I_2 = I_3 + I_4$$

✔ *Quick check 1, 2*

Series and parallel

- Connected *in series*: current flows through components one after the other.
- Connected *in parallel*: current divides up and is shared between components. It can help to show two currents flowing, one for each component.

▶▶ *Resistances in series and parallel – see pages 30–31.*

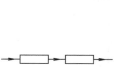

in series

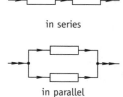

in parallel

Coulombs and amps

Charge Q is measured in **coulombs** (C).

Current I is measured in amperes, **amps** (A).

One amp is one coulomb per second: **$1\ A = 1\ C\ s^{-1}$**.

One coulomb is the charge which passes when a current of 1 amp flows for 1 second.

▶▶ *More about the definition of the ampere on page 44.*

Formulae relating Q, I and t

To calculate the current I flowing when charge ΔQ passes a point in time interval Δt:

$$I = \frac{\Delta Q}{\Delta t}$$

This reminds us that current is a *rate* of flow – that's why we divide by time.

To calculate the charge:

$$\Delta Q = I \times \Delta t$$

The bigger the current, and the longer it flows, the more the charge that passes.

Worked examples

1 What current flows when a charge of 600 C passes a point in 1 minute?

   Step 1 Write down what you know, and what you want to know:

   $$\Delta Q = 600 \text{ C}, \Delta t = 60 \text{ s}, I = ?$$

   Step 2 Choose the appropriate equation, substitute and solve:

   $$I = \frac{\Delta Q}{\Delta t} = \frac{600 \text{ C}}{60 \text{ s}} = 10 \text{ A}$$

2 How much charge passes a point when a current of 10 mA flows for 10 s?

   Step 1 Write down what you know, and what you want to know:

   $$I = 10 \text{ mA (or } 10^{-2} \text{ A or 0.01 A)}, \Delta t = 10 \text{ s}, \Delta Q = ?$$

   Step 2 Choose the appropriate equation, substitute and solve:

   $$\Delta Q = I \times \Delta t$$

   $$= 10 \text{ mA} \times 10 \text{ s} = 100 \text{ mC}$$

❖ Use SI units.

❖ Don't forget units. Strictly speaking, this is the *average* current flowing in this time.

❖ Since we are working in mA (milliamps), the answer is in mC (millicoulombs). We could have converted from mA to A.

✓ *Quick check 3, 4*

? *Quick check questions*

1 Calculate current I_4, as shown in the figure.
2 Currents of 2.5 A, 1.0 A and 10.5 A are supplied by a car battery. What is the total current it supplies?
3 What current is flowing if 240 mC of charge flows past a point in 30 s?
4 A motorist is having trouble getting his car to start. The battery supplies a current of 100 A for 1 minute. How much charge flows from the battery in this time?

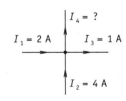

$I_4 = ?$

$I_1 = 2 \text{ A}$ $I_3 = 1 \text{ A}$

$I_2 = 4 \text{ A}$

Resistance

A **potential difference** (p.d.) is needed to push a current through a component. The **electrical resistance** of the component tells us how easy (or rather, how difficult) it is to make current flow through it.

> **The greater the resistance, the smaller the current that flows for a given p.d.**

▶▶ *For the meaning of potential difference (p.d.) see page 34.*

Defining resistance

The **resistance** (R) of a component is the ratio of the p.d. (V) *across* it to the current (I) flowing *through* it. It is defined by the equation:

$$\text{resistance} = \frac{\text{p.d.}}{\text{current}} \qquad R = \frac{V}{I}$$

The equation for resistance can be rearranged:

$$V = IR \qquad I = \frac{V}{R}$$

Ohms, amps and volts

Resistance is measured in **ohms** (Ω).

One ohm is one volt per amp: $\mathbf{1\ \Omega = 1\ V\ A^{-1}}$.

So it takes a p.d. of 1 V to make a current of 1 A flow through a 1 Ω resistor, and it takes a p.d. of 10 V to make a current of 1 A flow through a 10 Ω resistor.

- 1 kilohm = 1 kΩ = 10^3 Ω = 1000 Ω
- 1 megohm = 1 MΩ = 10^6 Ω = 1000 000 Ω

C The symbol Ω is the Greek letter omega.

✓ *Quick check 1, 2*

Resistors in series

When resistors are connected **in series** (end-to-end), the current flows through one and then through the next, and so on.

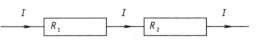

- Resistors in series must have the *same current* flowing through them.
- The p.d. of the supply is *shared* between them.

To find the combined resistance R of two or more resistors in series, add up their individual resistances:

$$R = R_1 + R_2 + R_3 + \ldots \text{ in series}$$

Worked example

A 20 Ω resistor and a 5 Ω resistor are connected in series with a 10 V battery. What is the p.d. across each resistor?

Step 1 Sketch a diagram, and mark on it the available information.

Step 2 Calculate the combined resistance:

$$R = R_1 + R_2 = 20 \ \Omega + 5 \ \Omega = 25 \ \Omega$$

Step 3 Calculate the current that flows:

$$I = \frac{V}{R} = \frac{10 \ V}{25 \ \Omega} = 0.4 \ A$$

Step 4 Calculate the p.d. across each resistor:

Across 20 Ω: $V = IR = 0.4 \ A \times 20 \ \Omega = 8 \ V$

Across 5 Ω: $V = IR = 0.4 \ A \times 5 \ \Omega = 2 \ V$

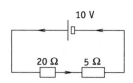

A useful rule: The bigger resistor gets a bigger share of the p.d.

✓ *Quick check 3*

Resistors in parallel

When resistors are connected **in parallel** (side-by-side), the *current* divides up, part of it flowing through each resistor.

- Resistors in parallel have the *same p.d.* across them.
- The current flowing from the supply is *shared* between them.

Since $I = V/R$ for *each* resistor (see page 30), to find the combined resistance R of two or more resistors in parallel, add up the *reciprocals* of their individual resistances:

$$\frac{1}{R} = \frac{1}{R_1} + \frac{1}{R_2} + \frac{1}{R_3} + \dots \textbf{ in parallel}$$

Worked example

A 20 Ω resistor and a 5 Ω resistor are connected in parallel with a 10 V battery. What current flows from the battery?

Step 1 Sketch a diagram, and mark on it the available information.

Step 2 Calculate (in two stages!) the combined resistance:

$$\frac{1}{R} = \frac{1}{R_1} + \frac{1}{R_2} = \frac{1}{20 \ \Omega} + \frac{1}{5 \ \Omega} = 0.05 \ \Omega^{-1} + 0.20 \ \Omega^{-1} = 0.25 \ \Omega^{-1}$$

$$R = \frac{1}{0.25 \ \Omega^{-1}} = 4 \ \Omega \ (R \text{ is always less than the smallest of } R_1, R_2, \text{ etc.})$$

Step 3 Calculate the current from the combined resistance and the p.d.:

$$I = \frac{V}{R} = \frac{10 \ V}{4 \ \Omega} = 2.5 \ A$$

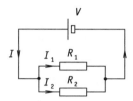

Another method: Calculate the current through each resistor separately, and add them together.

✓ *Quick check 4, 5*

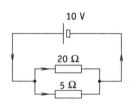

? *Quick check questions*

1 What is the resistance of a resistor if a p.d. of 10 V makes a current of 2 A flow through it?

2 What p.d. will make a current of 20 mA flow through a 500 kΩ resistor?

3 Two 20 Ω resistors are connected in series with a 5 V supply. What is the p.d. across each resistor?

4 What is the resistance of 20 Ω, 30 Ω, 60 Ω resistors connected in parallel?

5 Three 2 kΩ resistors are connected in parallel across a 12 V supply. What current flows from the supply?

Recall that an ohm is a volt per amp.

How is the p.d. shared between resistors in series?

First calculate the current that flows through one resistor.

Ohm's law

A potential difference (p.d.), or **voltage**, pushes current through a conductor. The greater the p.d., the greater the current. An **ohmic conductor** is one in which the current that flows is proportional to the p.d. that pushes it. It obeys **Ohm's law**.

Remembering the equation $V = IR$ (page 30), this means that the resistance of the conductor is constant, and does not depend on the p.d. across it.

Measuring resistance

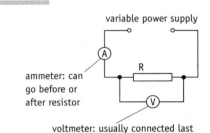

- An **ammeter** is an instrument for measuring current, for example through a resistor. It is connected *in series* with the resistor.

- A **voltmeter** is an instrument for measuring the p.d. (voltage) across a resistor. It is connected *in parallel* with the resistor.

- Reversing the connections to the resistor makes the current flow through it in the opposite direction. This gives negative values of current I and voltage V.

- The results can be plotted as a **current–voltage characteristic** graph, or I–V graph. The figure shows an I–V graph for a metallic conductor at constant temperature. An ohmic conductor gives a straight line through the origin.

It is usual to plot p.d. on the x-axis.

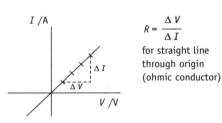

$$R = \frac{\Delta V}{\Delta I}$$

for straight line through origin (ohmic conductor)

✓ *Quick check 1*

Non-ohmic conductors

Metals obey Ohm's law – they are ohmic conductors – provided the temperature remains constant. A **non-ohmic conductor** has an I–V characteristic graph which is not a straight line. The figure shows two examples.

Filament lamp: As the metal filament gets hotter with increasing current and voltage, its resistance increases. The current is less than it would be if it remained proportional to the voltage.

Semiconductor diode: In the forward direction, allows current to flow when the p.d. is above 0.7 V. In the reverse direction, only a very small current flows.

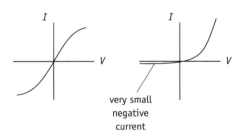

Filament lamp   **Semiconductor diode**

very small negative current

✓ *Quick check 2*

Temperature dependence

The figure shows two examples.

- *Metal:* Resistance increases gradually as temperature is increased. (Atomic vibrations increase, so conduction electrons are scattered more.)

- *NTC thermistor:* Resistance decreases rapidly over a narrow range of temperature.

NTC means 'negative temperature coefficient': hotter = less resistance.

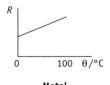

Metal

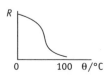

NTC thermistor

Resistivity

Some materials resist the flow of electric current more than others. The property that describes this is **resistivity**, ρ (Greek letter rho).

To calculate the resistance R of a wire, for example, we need to know three things:

- its length l – the longer the wire, the greater its resistance: $R \propto l$
- its cross-sectional area A – the fatter the wire, the less its resistance: $R \propto 1/A$
- the resistivity of the material ρ: $R \propto \rho$ (ρ is a constant for a given material, at a given temperature). The formula is

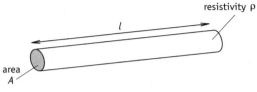

$$R = \frac{\rho l}{A} \quad \text{or} \quad \rho = \frac{RA}{l}$$

Resistivity is measured in Ω **m (ohm metres).**

A typical value for a good conductor: resistivity of copper = 1.6×10^{-8} Ω m.

◖ Take care – not ohms per metre!

Worked example

What is the resistance of a 20 m length of silver wire of diameter 1 mm? (Resistivity of silver = 1.6×10^{-8} Ω m.)

Step 1 Write down what you know, and what you want to know; you will have to calculate A.

l = 20 m, ρ = 1.6×10^{-8} Ω m, $A = \pi r^2 = \pi \times (0.5 \times 10^{-3}$ m$)^2 = 7.85 \times 10^{-7}$ m^2, R = ?

◖ Remember to halve the diameter to find the radius.

Step 2 Calculate R.

$$R = \frac{\rho l}{A} = \frac{1.6 \times 10^{-8}\,\Omega\,\text{m} \times 20\,\text{m}}{7.85 \times 10^{-7}\,\text{m}^2} = 0.41\ \Omega$$

✓ *Quick check 3, 4*

1 The table shows experimental results for a carbon resistor of resistance R. Plot a graph and use it to deduce R.

current I/mA	180	340	550	700	910
p.d. V/V	2.0	4.0	6.0	8.0	10.0

2 Use the graph in the figure to decide at what voltage the filament lamp and the resistor have the same resistance R. What is the value of R?

3 What is the resistance of a 5 m length of copper wire of cross-sectional area 1 mm^2? (Resistivity of copper = 1.6×10^{-8} Ω m.)

4 Does the resistivity of a metal increase or decrease with temperature?

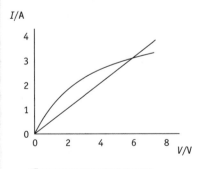

◖ 1 mm^2 = 10^{-6} m^2.

p.d. and e.m.f.

We use electricity to transfer energy from place to place. If you use a battery to light a bulb, electricity transfers energy from the battery to the bulb. The *voltage* of a source of electricity is a measure of these energy transfers.

We use two (more correct) terms for voltage:

- **potential difference**, p.d., symbol *V*
- **electromotive force**, e.m.f., symbol *E*

❶ Take care always to talk about the potential difference *across* a component, or *between* two points. A p.d. does not 'go through' a component.

Around a circuit

In this circuit, the cell is pushing a current (a flow of charge) through the lamp.

The current is the same at all points around the circuit.

- *Inside the cell:* Charges flow through the cell, collecting energy.
- *Inside the lamp:* The same charges flow through the lamp, giving up energy.

The voltmeters give equal but opposite readings (positive and negative).

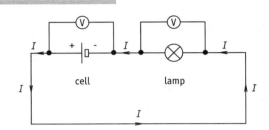

❶ Note that the current inside the cell flows from negative to positive – see page 28.

The meaning of e.m.f.

The voltage shown on a cell or battery tells you its **e.m.f.** From this, you can tell the amount of energy given to each coulomb of charge which passes through.

> **The e.m.f. of a supply is the work it does in pushing 1 C of charge around a complete circuit.**

- A 1.5 V cell gives 1.5 J to each coulomb.
- A 6 V battery gives 6 J to each coulomb.
- The 230 V mains gives 230 J to each coulomb.

> **Similarly, the p.d. across a component (such as a resistor or lamp) tells you how many joules of energy are given up by each coulomb passing through.**

✓ *Quick check 1*

Formulae and units

Both p.d. *V* and e.m.f. *E* are measured in volts (V).

Energy *W* is measured in joules (J).

Charge *Q* is measured in coulombs (C).

Energy transferred = voltage × charge: $W = VQ$.

One volt is one joule per coulomb: $1 \text{ V} = 1 \text{ J C}^{-1}$.

❶ Think of *W* for work. Avoid using *E* for energy.

✓ *Quick check 2, 3*

Combining e.m.f.s

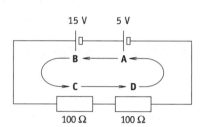

$E = 1.5$ V $E = 4 \times 1.5$ V $= 6$ V $E = 3 \times 1.5$ V $- 1.5$ V $= 3$ V

Kirchhoff's second law

Trace the movement of 1 C of charge around the circuit.

- At **A**: 5 J gained
- At **B**: 15 J gained; total energy gained = 20 J
- At **C**: 10 J lost
- At **D**: 10 J lost; total energy lost = 20 J

By the time the charge has completed the circuit, it has lost as much energy as it gained. This is an example of **conservation of energy**. We can think of this in terms of voltages around the circuit:

> ***Kirchhoff's second law*:**
> **The sum of the e.m.f.s around any circuit loop is equal to the sum of the p.d.s: $\Sigma E = \Sigma R$.**

✓ *Quick check 4–6*

? Quick check questions

1. How much energy is transferred to each coulomb of charge by a 9 V battery?

2. 10 C of charge flows through a p.d. of 6 V. How much energy is transferred?

3. A current of 2.5 A flows through a resistor for 1 minute. It transfers 600 J of energy to the resistor. What is the p.d. across the resistor?

4. Two resistors are connected in series with the 230 V mains supply. The p.d. across one resistor is 70 V. What is the p.d. across the other?

5. Calculate:
   a   the total e.m.f. in this circuit;
   b   the current which flows around it;
   c   the p.d. across each resistor.
   Show that Kirchhoff's second law is satisfied.

6. How much work is done by the 230 V mains supply in pushing 1 C of charge round a circuit?

▶ Start by calculating how much charge flows.

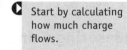

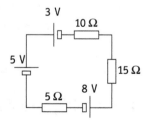

Internal resistance and potential dividers

We use a cell or power supply to provide a p.d. Sometimes it provides less than we expect, because of its **internal resistance**. Sometimes we want a smaller p.d. than it supplies, and we use a **potential divider** to reduce the p.d.

Internal resistance

Think about the current flowing around a circuit, pushed by a cell or power supply. The same current flows all the way round. In particular, it flows through the *interior* of the supply (from negative to positive). The interior of a supply is made up of chemicals or metal wire, and must have resistance. This is the internal resistance r of the supply.

We show the cell as a 'perfect cell', marked with its e.m.f. E, and a small resistor r in series with it. The current I flows through a combined resistance $R + r$.

$$E = I\,(R + r)$$
$$E - Ir = IR$$

e.m.f. – lost volts = terminal p.d. of cell

We get fewer volts out of the supply because some of the e.m.f. is used up in overcoming the internal resistance when a current is flowing.

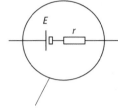

the circle is optional; it indicates that E and r are part of the same thing

▶▶ *The meaning of e.m.f. is explained on page 34.*

Measuring E and r

Varying the value of R makes the current I change. The graph shows that the greater the current that flows from the supply, the smaller its terminal p.d. The graph is roughly a straight line:

- gradient = $-r$,
- intercept on y-axis = E.

intercept = E gradient = $-r$

variable resistor

Using a high-resistance voltmeter

A digital voltmeter has a resistance of millions of ohms. Connect it across a supply, and only a tiny (negligible) current will flow. Its reading will therefore indicate the e.m.f. of the supply ('lost volts' = 0).

The e.m.f. of a supply is the p.d. across its terminals when no current flows.

✔ *Quick check 1, 2*

Worked example

A power supply of e.m.f. 6 V and internal resistance 2 Ω is connected across a 10 Ω resistor. What current flows through the resistor, and what is the terminal p.d. of the supply?

Step 1 Draw a diagram, showing both R and r.

Step 2 Calculate the total resistance in the circuit.

$$R + r = 10\ \Omega + 2\ \Omega = 12\ \Omega$$

Step 3 Calculate the current that flows.

$$I = \frac{E}{R + r} = \frac{6\ V}{12\ \Omega} = 0.5\ A$$

Step 4 Calculate the terminal p.d.

$$V = IR = 0.5\ A \times 10\ \Omega = 5\ V$$

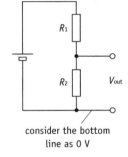

❶ 1 V has been 'lost' in overcoming the internal resistance.

✓ *Quick check 3*

Potential dividers

Reduce the p.d. provided by a supply by connecting two resistors across its terminals. Tap off the required p.d. V_{out} from the point between them.

The bigger resistor takes the bigger share of the p.d.

Use equal resistors to give *half* the p.d. of the supply.

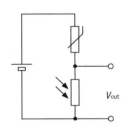

consider the bottom line as 0 V

Varying resistances

- Shining light on the LDR (light-dependent resistor) will decrease its resistance. V_{out} will decrease.

- Warming the NTC thermistor will decrease its resistance. V_{out} will decrease.

Swop the resistors over if you want V_{out} to increase.

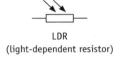

LDR
(light-dependent resistor)

NTC thermistor

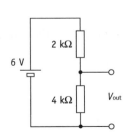

✓ *Quick check 3, 4*

❓ *Quick check questions*

1 A power supply of e.m.f. 500 V and internal resistance 0.1 Ω is connected to a heater of resistance 125 Ω. What current flows through the heater?

2 A high resistance voltmeter is connected across a cell, and gives a reading of 1.55 V. A 600 Ω resistor is added in parallel with the voltmeter, which now reads 1.50 V. What are the e.m.f. and internal resistance of the cell?

3 Calculate V_{out} as shown in the figure. Notice that one resistor has twice the resistance of the other.

4 A potential divider is constructed from a 50 Ω fixed resistor and a thermistor whose resistance changes from 450 Ω at 20°C to 50 Ω at 80°C. The divider is connected across a 10 V supply. Draw a diagram of this arrangement, and show that its output voltage will vary between 9 V and 5 V as the temperature is varied between 20°C and 80°C.

Electrical power

We use electricity as a convenient way of transferring energy from place to place. *Voltage* is a measure of how much energy is transferred to or from each coulomb of charge. *Current* tells us about the rate at which charge moves. Combining these quantities tells us about the *rate* at which energy is transferred by a current – the **electrical power**.

Power in general

We met the general idea of *power* on page 17. Power is the rate at which energy is transferred.

$$\text{power} = \frac{\text{energy transferred}}{\text{time taken}} \qquad P = \frac{W}{t}$$

$$\text{energy transferred} = \text{power} \times \text{time} \qquad W = Pt$$

Units

Power is measured in **watts**, W.

1 watt = 1 joule per second $1\ \text{W} = 1\ \text{J s}^{-1}$

1 kilowatt = 1 kW = 10^3 W = 1000 W

1 megawatt = 1 MW = 10^6 W = 1000 000 W

1 gigawatt = 1 GW = 10^9 W = 1000 000 000 W

▶▶ *Mechanical power is the rate at which energy is transferred by a force – see page 17.*

✓ *Quick check 1, 2*

Calculating electrical power

The greater the current I and the greater the p.d. V that it is flowing through, the greater the power P.

$$\text{power} = \text{current} \times \text{p.d.} \qquad P = IV$$

Combining this equation with $V = IR$ gives two alternative forms:

$$P = I^2R \quad \text{and} \quad P = \frac{V^2}{R}$$

C You may find it easier to remember this as watts = amps × volts.

C Choose the form of equation according to the information you have in any question.

Worked example

A car headlamp bulb is labelled '12 V, 48 W'. What is its resistance in normal operation? (The label indicates its normal operating voltage and power.)

Step 1 Write down what you know, and what you want to know:

$$V = 12 \text{ V}, \quad P = 48 \text{ W}, \quad R = ?$$

Step 2 Select the appropriate equation and rearrange it to make R the subject:

$$P = \frac{V^2}{R} \quad \text{so} \quad R = \frac{V^2}{P}$$

Step 3 Substitute values and calculate the answer:

$$R = \frac{(12 \text{ V})^2}{48 \text{ W}} = 3 \ \Omega$$

✓ *Quick check 3, 4*

Another energy unit – the kilowatt-hour (kWh)

A 1 kW heater used for 1 h transfers 1 kWh of energy.

A 2 kW heater used for 5 h transfers 10 kWh of energy.

Energy transferred (kWh) = power (W) × time (h)

1 kWh = 3600 000 J = 3.6 MJ, because 1 kW = 1000 J s^{-1} and 1 h = 3600 s.

▶ A joule is a small amount of energy. It's often easier to work in kilowatt-hours.

✓ *Quick check 5*

SI unit summary

It is often easier to learn equations linking units rather than quantities.

quantity	unit	equivalents	in words
current I	ampere, A	1 A = 1 C s^{-1}	amp = coulomb per second
p.d. V, e.m.f. E	volt, V	1 V = 1 J C^{-1}	volt = joule per coulomb
resistance R	ohm, Ω	1 Ω = 1 V A^{-1}	ohm = volt per amp
power P	watt, W	1 W = 1 J s^{-1}	watt = joule per second

? *Quick check questions*

1 The output of a power station is stated as 450 MW. How many joules of electrical energy does it supply each second?

2 A battery transfers 30 J of energy each minute. What power is this?

3 An electric motor draws a current of 2.5 A from a 12 V supply. What power does it transfer?

4 A 60 W lamp has a resistance of 2.4 Ω. What current flows through it in normal operation?

5 A 150 W light bulb is left on for 24 hours. How many kWh of energy are transferred in this time?

Magnetic fields

Magnetic fields can be produced in two ways:

- by permanent magnets,
- by electric currents.

In both cases, electric charges are moving, usually electrons. The Earth's magnetic field is thought to be created by strong electric currents flowing in its iron core.

Permanent magnets

A magnet's **north pole** is so called because it is attracted towards the *geographical* north. (Take care! This means that there must be a *magnetic* south pole at the Earth's geographical north pole, because north poles of magnets are attracted to south poles.)

Representing the field

We represent a magnetic field by **field lines** (**flux lines**). These come out of north poles, and go into south poles.

- Lines close together: strong field – high **flux density**.
- Lines far apart: weak field – low flux density.

▶▶ *Flux density tells us how strong the field is – page 42. It is measured in teslas – page 45.*

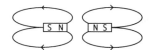

Attraction and repulsion

Like poles repel. Unlike poles attract.

Electromagnets: solenoid

A coil (or **solenoid**) carrying a current acts like a bar magnet. Its magnetic field has the same shape as that of a bar magnet. One end is a north pole, the other is a south pole.

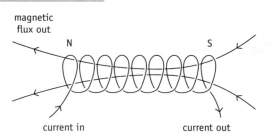

magnetic flux out

current in   current out

> Imagine curling the fingers of your *right* hand round the solenoid so that they follow the current around the coil. Your thumb points in the direction of the flux emerging from the solenoid. This is the **right-hand grip rule.**

Use the right-hand grip rule to find the direction of the flux emerging from inside the coil.

To increase the flux density:

- increase the current,
- use more turns of wire in the same space,
- add an iron core.

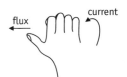

flux   current

C Note that the flux inside the solenoid flows from the south pole to the north pole.

The figure shows another rule for finding the poles of a solenoid.

Look at one end. If the current is going anticlockwise, you are at the N end and the flux is flowing towards you. If clockwise, you are at the S end and it is flowing away from you.

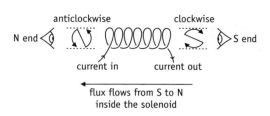

✓ Quick check 1

A flat circular coil

This is similar to a solenoid. The lines of flux are continuous loops. (Picture a solenoid squashed up.)

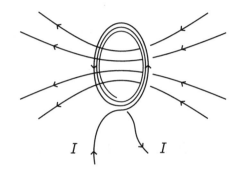

A long straight wire

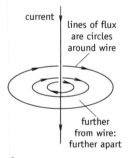

This is easier to draw as a plan view.

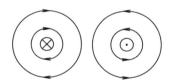

The **corkscrew/pencil sharpener rule** says that if the current is flowing away from you, the flux goes clockwise, like turning a corkscrew into a cork.

✓ Quick check 2–4

? Quick check questions

1 Will the magnet and solenoid attract or repel?

2 Draw a plan view to show the pattern of magnetic flux around a long straight wire when the current is flowing vertically upwards, out of the paper.

3 Modify your answer to **2** to show how the flux pattern will change if the current flowing in the wire is doubled.

4 How do your diagrams (questions **2** and **3**) show that the magnetic field is weaker the further you are from the wire?

Magnetic forces

When an electric current flows *across* a magnetic field, it feels a force. A current-carrying conductor is pushed by this force. This is made use of in electric motors.

Force on a current-carrying conductor

The current must flow *across* the lines of flux.
(Take care – if no current flows, there is no force.)

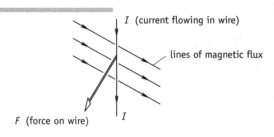

I (current flowing in wire)

lines of magnetic flux

F (force on wire)   *I*

Fleming's left-hand rule

This gives the relative directions of current, field and force (i.e. *motion*). All three are at right angles to each other. Remember to use your *left* hand.

thuMb for Motion (force)
First finger for Field
seCond finger for Current

✓ *Quick check 1, 2*

Calculating the force

The force *F* in newtons (N) is proportional to three things:

- *B*, magnetic flux density in **teslas** (T) – see page 45,
- *I*, current in amps (A),
- *l*, length of wire in field in metres (m).

The formula is simply

$$F = BIl$$

This relationship applies only when the current is at 90° to the magnetic flux. At other angles, the force is less. In fact, if the current is at θ° to the flux direction,

$$F = BIl \sin\theta$$

▶▶ *Charges moving in a magnetic field – page 44.*

Worked example

A current of 10 A flows in a wire which crosses a magnetic field of flux density 50 mT at 90°. What force acts on each metre length of the wire?

Step 1   Write down what you know, and what you want to know:

$$I = 10 \text{ A}, \quad B = 50 \text{ mT} = 50 \times 10^{-3} \text{ T}, \quad l = 1 \text{ m}, \quad F = ?$$

Step 2   Write down the equation, substitute and solve:

$$F = BIl = 50 \times 10^{-3} \text{ T} \times 10 \text{ A} \times 1 \text{ m} = 0.5 \text{ N}$$

 ✓ *Quick check 3, 4*

Explaining the force

The force comes about because the magnetic flux resulting from the current in the wire interacts with the flux of the external magnetic field.

Using the force

An electric motor consists of:

- a coil of wire carrying an electric current – this produces a magnetic field;

- permanent magnets or electromagnets – these produce an external magnetic field.

The two fields interact to make the coil turn.

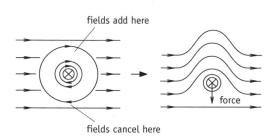

fields add here

fields cancel here

force

? Quick check questions

1 This wire lies in a magnetic field which is directed downwards into the paper. Copy the diagram and add an arrow to show the direction of the magnetic force on the wire.

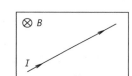

2 This coil of wire lies in a magnetic field, as shown. Which sides will experience *no* magnetic force? In which direction will the coil start to turn (clockwise or anticlockwise)?

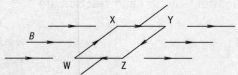

3 A wire of length 10 m carries a current of 4 A. It lies at right angles to a magnetic field. If the force on the wire is 2 N, what is the flux density of the field?

4 Two wires lie parallel to one another in a uniform magnetic field. One is 4.0 cm long and carries a current of 8.2 A; the other is 6.4 cm long and carries 6.0 A. Each feels a force due to the magnetic field. Which will experience the greater magnetic force?

The ampere and the tesla

The SI system of units consists of certain *fundamental units* (m, kg, s, etc.) and others which are derived from these. The *ampere* is a fundamental unit; the *tesla* is a derived unit.

By looking at equations which relate quantities to one another, we can see how their units are related.

Some fundamental SI units

quantity	unit	abbreviation
mass	kilogram	kg
length	metre	m
time	second	s
electric current	ampere, amp	A

Defining the ampere

The definition of the ampere is based on the following situation.

Two long, parallel wires each carry an electric current. Each current produces its own magnetic field. The fields interact, as shown.

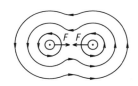

The bigger the currents, the bigger the force between the wires. By considering standard wires separated by a standard distance, this provides a way of standardising the ampere. The definition of the ampere is thus based on measurements of the force between two parallel current-carrying wires.

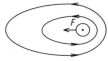
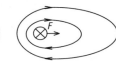

Note that each current feels the *same* force, even if one current is bigger than the other. This is an example of Newton's third law of motion – action and reaction are equal and opposite.

✓ *Quick check 1, 2*

Why two parallel currents attract

Here are two ways of showing why this happens.

- The magnetic fields between the wires are in opposite directions, so they cancel.

- The current in wire 1 exerts a force *BIl* (see page 42) on the current in wire 2 (and vice versa).

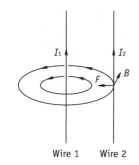

Wire 1   Wire 2

Units derived from the ampere

To find the unit of magnetic flux density B, we have to rearrange the equation $F = BIl$ (page 42):

$$B = \frac{F}{Il}$$

So the unit of B is the same as the unit of F/Il. It is called the **tesla** and is defined by:

> $$1\ \text{T} = 1\ \text{N A}^{-1}\ \text{m}^{-1}$$

> **1 T is the flux density which, acting perpendicular to a current of 1 A flowing in a 1 m length of conductor, produces a force on the current of 1 N.**

There is no need to remember this equation, but you should recall the equation for F.

derived unit	quantity equation	unit equation
coulomb, C	$Q = It$	$1\ \text{C} = 1\ \text{A s}$
volt, V	$V = \dfrac{W}{Q}$	$1\ \text{V} = 1\ \text{J C}^{-1}$
tesla, T	$B = \dfrac{F}{Il}$	$1\ \text{T} = 1\ \text{N A}^{-1}\ \text{m}^{-1}$

✓ *Quick check 3*

❓ Quick check questions

1 The diagram shows the magnetic fields around two current-carrying wires, perpendicular to the page. What can you say about the directions of the two currents? What does the diagram show about the forces on the two wires?

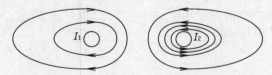

2 The diagram in question **1** shows two current-carrying wires 1 and 2, and how the force on wire 1 arises. Draw a diagram to show how the current in wire 1 exerts a force BIl on the current in wire 2.

3 The diagram shows how some SI units are related. Copy the diagram and, next to each line, add an equation which relates the units.

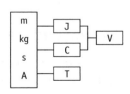

The electromagnetic spectrum

A prism or diffraction grating can split white light up into the familiar **spectrum**, from red to violet. This spectrum shows all the wavelengths present in visible light, arranged from the longest wavelength (red) to the shortest (violet).

This is only part of a much wider spectrum of *electromagnetic radiation*.

The visible spectrum

The symbol for wavelength is the Greek letter λ (lambda).

The range of wavelengths of visible light is from 400 nm (violet) to 700 nm (red).

400 nm (nanometres) = 400×10^{-9} m = 4×10^{-7} m

It is useful to remember a typical value for the wavelength of visible light – say, λ = 500 nm.

> These are only approximate values – it depends on your eyes!

✓ *Quick check 1*

Beyond the visible

region of spectrum	range of wavelengths (shortest to longest)
gamma rays (γ-rays)	10^{-16} to 10^{-10} m
X-rays	10^{-13} to 10^{-8} m
ultraviolet	10^{-8} m to 4×10^{-7} m (400 nm)
visible	400 nm to 700 nm
infrared	7×10^{-7} m (700 nm) to 10^{-3} m
microwaves	10^{-3} m to 10^{-1} m
radio waves	10^{-1} m to 10^{6} m or more

The electromagnetic spectrum includes wavelengths ranging over many orders of magnitude. It is divided into different regions, but the boundaries between them are not well defined.

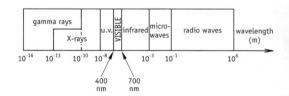

You need to learn:

- the names of the different regions,
- the order they appear in,
- their approximate wavelength ranges.

✓ *Quick check 2*

Energy and wavelength

Gamma rays have the shortest wavelengths and highest energies. Radio waves of long wavelength have the lowest energies.

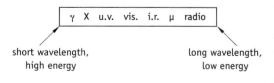

▶▶ *Frequency is related to wavelength – see pages 48 and 66.*

✓ *Quick check 3*

Speed of electromagnetic radiation

All types of electromagnetic radiation travel at the same speed through free space. This is often referred to as the **speed of light**, symbol c.

In the SI system of units, the value of c is defined as

$$c = 299\ 792\ 458\ \text{m s}^{-1}$$

This defines how metres and seconds are related in the SI system.

For many calculations, we can use an approximate value:

$$c \simeq 3 \times 10^8\ \text{m s}^{-1}$$

Try to remember this approximate value.

> **O** 'Free space' means completely empty space – a perfect vacuum. Any material will slow electromagnetic radiation down.

✓ Quick check 4, 5

How electromagnetic radiation is produced

Gamma rays (γ-rays)	By radioactive substances. A γ-ray may be emitted from the nucleus of a radioactive atom when it decays.
X-rays	By fast-moving (energetic) electrons when they are slowed down. Their energy is transformed to one or more X-rays.
ultraviolet	By very hot objects.
visible light	By hot objects. An electron drops to a lower energy level in an atom.
infrared	By all objects, so long as they are warmer than absolute zero.
microwaves, radio waves	By oscillating (vibrating) electrons, e.g. an alternating current in a radio or TV transmitter aerial.

> **O** X-rays and γ-rays are very similar – we give them different names according to how they are produced.

❓ Quick check questions

1 The red limit of the visible spectrum is at about 700 nm. Express this in standard form, i.e. scientific notation, using powers of 10.

2 In what region of the electromagnetic spectrum does each of the following wavelengths lie?   1 km, 800 nm, 500 nm, 1 nm

3 Which are the most and least energetic regions of the electromagnetic spectrum?

4 Which of the following give a good approximation to the speed of light in free space? $300\ 000\ \text{km s}^{-1}$, $300\ 000\ 000\ \text{m s}^{-1}$, $300 \times 10^6\ \text{m s}^{-1}$

5 Roughly how many seconds does it take light to travel from the Sun to the Earth, a distance of 150 million km?

> **O** Try to answer question 2 without looking at the table of wavelengths.

Photons

Sometimes we think of light as being made up of 'particles' called **photons**, each with its own energy; sometimes we think of light as being made up of electromagnetic waves, which have a frequency and wavelength.

Wave frequency

The **frequency** of a wave is the number of wavecrests passing any given point per second. Its units are therefore s^{-1}, otherwise known as **hertz** (Hz). The longer the wavelength, the lower the frequency.

▶▶ *For more details of this relationship, see pages 64 and 66.*

Energy and frequency

The energy of a photon is related to the wave frequency f by

photon energy = hf

where h is a constant known as the **Planck constant**: $h = 6.63 \times 10^{-34}$ J s.

You can think of the units J s as 'joules per hertz'. The more hertz (the higher the frequency), the more joules are carried by each photon.

This equation links a particle property (photon energy) to a wave property (frequency).

▶▶ *More about particle and wave pictures – page 50.*

✓ *Quick check 1*

The electronvolt – an energy unit

Photon energies are very small. A more convenient unit than the joule is the **electronvolt** (eV).

1 eV is the energy transferred when an electron moves between two points separated by a p.d. of 1 V.

$$1 \text{ eV} = 1.6 \times 10^{-19} \text{ J}$$

- To convert from J to eV: divide by 1.6×10^{-19} (i.e. multiply by 6.25×10^{18}).
- To convert from eV to J: multiply by 1.6×10^{-19}.

▶▶ *Remember the equation W = QV (energy = charge × voltage) on page 34. That's where the definition of the electronvolt comes from. The electron charge is e = -1.6×10^{-19} C.*

✓ *Quick check 2–4*

The de Broglie equation

We have already seen one equation which links a wave property (frequency f) with a particle property (photon energy E):

$$E = hf$$

The **de Broglie equation** is another equation which allows us to translate between wave behaviour and particle behaviour. It links wavelength λ (lambda) with particle momentum p. The more momentum a particle has, the shorter its wavelength:

$$\lambda = \frac{h}{p}$$

Note that, in both of these equations, the Planck constant h connects the wave quantity to the particle quantity.

A note on momentum

You can calculate your momentum by multiplying your mass by your velocity, $m \times v$. The greater the mass of a particle, and the faster it moves, the greater its momentum. The formula $m \times v$ doesn't apply to very fast-moving particles with speeds approaching the speed of light, so it is better to use the symbol p for momentum. The units of momentum are kg m s^{-1} or N s.

▶▶ *Lots more about momentum in Module D of the companion book – Revise A2 Physics for OCR, Specification A.*

Worked example

An electron has momentum 2×10^{-24} kg m s^{-1}. What is its wavelength?

$$\lambda = \frac{h}{p} = \frac{6.63 \times 10^{-34} \text{ J s}}{2 \times 10^{-24} \text{ kg m s}^{-1}} = 3.31 \times 10^{-10} \text{ m}$$

(This value of momentum is for an electron moving at roughly 2×10^{6} m s^{-1}. Its wavelength is similar to the size of an atom, which is why such electrons are diffracted by crystalline graphite – see page 52.)

✓ *Quick check 5*

The de Broglie equation applies to *all* particles, no matter how big. If you run, you are like a moving particle. Your momentum might be 500 kg m s^{-1}; your wavelength would then be about 10^{-32} m. This is much too small for you to observe any wave effects, such as diffraction or interference.

? *Quick check questions*

(Planck constant $h = 6.63 \times 10^{-34}$ J s, speed of light in free space $= 3 \times 10^{8}$ m s^{-1}. See page 66.)

1 Calculate the energy of each photon in a beam of light of frequency 5×10^{14} Hz.

2 How many electronvolts of energy are transformed when an electron moves through a p.d. of 10.5 V?

3 What is the energy in eV of each photon in a beam of light of frequency 5×10^{14} Hz?

4 A laser produces photons of energy 1.90 eV. What is the wavelength of the laser light?

5 Light has momentum. Each photon in a beam of light carries an amount of momentum which can be calculated using the de Broglie equation. For light of wavelength 700 nm, what is the momentum of each photon?

◖ Use your answer to question 1 above.

The photoelectric effect

When light shines on certain metals, electrons break free. This is the **photoelectric effect.**

In order to explain this effect, Albert Einstein had to assume that, when light interacts with a metal, it behaves as particles (photons), not as waves. The energy of a photon of light is captured by a *conduction electron* in the metal, and the electron escapes from the surface of the metal.

Observing the photoelectric effect

When light is shone onto the photocell, a current starts to flow immediately. Even a very feeble light will work. The brighter the light, the greater the current.

The explanation is that the energy of the light helps conduction electrons to break free from the metal cathode. They cross to the anode; now there is a flow of charge all round the circuit. More light means more energy, so more electrons break free.

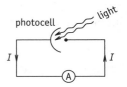

✓ *Quick check 1*

Explaining the photoelectric effect

Electrons don't normally escape from a metal. The conduction electrons are weakly held inside the metal. They need some energy to escape. Light can provide the necessary energy.

Why waves can't explain the effect

If we picture light waves falling on the metal, their energy is spread out all over the surface of the metal. It would take a long time for enough energy to be captured by the metal to free any electrons. The photoelectric effect is surprising because electrons break free as soon as the light is switched on.

Einstein argued that the energy of the light must be concentrated in tiny packets (photons). An individual conduction electron in the metal captures an individual photon; now the electron has enough energy to escape from the metal.

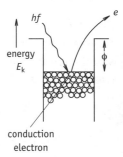

In the figure,

- hf is the energy of the photon,
- E_k is the kinetic energy of the electron,
- ϕ is the work function of the metal – the least amount of energy needed for an electron to escape from the metal.

How photons explain the effect

If a photon is captured by an electron in the metal, some of its energy is used to overcome the work function, and the rest ends up as the electron's kinetic energy.

The electrons which are highest up in the energy 'well' are the most energetic electrons. When one of these electrons captures a photon and escapes, it will have the maximum possible kinetic energy, $E_{k\,max}$. This gives us the equation:

$$hf = \phi + E_{k\,max}$$

Brighter light means more photons, so more electrons released. However, this doesn't increase the KE of the fastest electrons, because the individual photons do not have more energy. The current produced is proportional to the intensity of the light: greater intensity means more photons per second, so more electrons are released per second.

Threshold frequency

The frequency of the light must be above a certain minimum value, the **threshold frequency** $f_{threshold}$. Below this value, an individual photon does not have enough energy for an electron to overcome the work function. Hence:

$$hf_{threshold} = \phi$$

✓ *Quick check 2*

Worked example

A metal has a work function $\phi = 0.8$ eV. What is the greatest possible kinetic energy of an electron released by a photon of energy 1.4 eV?

Of the 1.4 eV of energy provided by the photon, 0.8 eV is used up in overcoming the work function. This leaves 0.6 eV of kinetic energy.

$$E_{kmax} = 1.4 \text{ eV} - 0.8 \text{ eV} = 0.6 \text{ eV}$$

(It is easiest to work in eV. However, you might have to calculate the photon energy from its frequency, and you might have to calculate the maximum speed of the electron using $\frac{1}{2} mv^2$.)

✓ *Quick check 3*

? Quick check questions

(Planck constant $h = 6.63 \times 10^{-34}$ J s, electron mass $m_e = 9.1 \times 10^{-31}$ kg, electron charge $e = 1.6 \times 10^{-19}$ C.)

1 A photocell makes a current flow around a circuit. Is the voltage across it a p.d. or an e.m.f.? Does the current *inside* the photocell flow from + to –, or from – to +?

2 What is the threshold frequency for a metal surface whose work function is 2.4×10^{-19} J?

3 Light of frequency 3.0×10^{14} Hz falls on a metal surface whose work function is 1.0 eV. What is the speed of the fastest electrons released?

❶ In this case, it is probably easiest to work in joules, rather than eV.

Wave–particle duality

For Einstein to explain the photoelectric effect, he had to assume that, when light interacts with the conduction electrons in a metal, it behaves as particles (photons). At other times, we know that light behaves as waves – for example, when it is diffracted (spread out) as it passes through a slit, or when two light waves interfere with one another. So light can behave as waves or as particles, depending on the circumstances. This is known as **wave–particle duality**.

In a similar way, particles (such as electrons) may also show wave-like behaviour.

Electron diffraction

Electrons can be diffracted. This shows that, when they pass through a fine grid, they behave like waves.

- A beam of fast-moving electrons is produced in a cathode ray tube.

- The electron beam passes through a thin layer of crystalline graphite (carbon).

- A diffraction pattern of fuzzy, light-and-dark rings is produced on a screen.

- To make the electrons go faster, increase the accelerating voltage. The diameter of the rings decreases. This shows that the wavelength decreases as the electrons go faster.

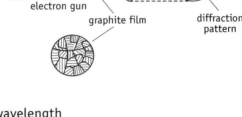

You can imagine how, as the wavelength of the waves gets smaller and smaller, the waves pass more easily through the gaps between the layers of carbon atoms, so they are diffracted less.

✓ *Quick check 1, 2*

Worked example

An electron accelerated through 100 V has momentum approximately equal to 5.4×10^{-24} kg m s^{-1}. Calculate the wavelength of such an electron and compare it to the separation of carbon atoms in graphite, approximately 0.37 nm.

Step 1   Use the de Broglie equation to calculate the electron's wavelength.

$$\lambda = \frac{h}{p} = \frac{6.63 \times 10^{-34} \text{ J s}}{5.4 \times 10^{-24} \text{ kg m s}^{-1}} = 1.2 \times 10^{-10} \text{ m}$$

Step 2   Compare with the separation of carbon atoms in graphite = 0.37 nm = 3.7×10^{-10} m.

The electron's de Broglie wavelength is about one-third of the atomic separation, so significant diffraction will be observed.

Waves or particles?

So, is light made up of waves or particles? Are electrons waves or particles? The answer is as follows.

We cannot say that light is waves, or particles. Sometimes light *behaves like* waves, sometimes it *behaves like* particles. And the same is true for electrons (and any other particle). We just have to learn when the wave picture gives the better explanation, and when the particle picture is better.

Waves and particles are things we are familiar with in our everyday lives. On the microscopic scale of electrons and photons, we discover that matter and radiation behave in a way that is a strange mixture of the two. We simply have to learn when to apply the wave picture or model, and when to think in terms of particles.

Some people try to invent funny little pictures of 'wavicles', half-wave and half-particle. This doesn't really help; it's best just to use either the wave model or the particle model, as appropriate.

✓ **Quick check 3**

? *Quick check questions*

1 In a diffraction experiment, what would happen to the speed of the electrons if the accelerating voltage was decreased? How would the diffraction pattern change?

2 In Einstein's explanation of the photoelectric effect, do the electrons behave as particles or as waves?

3 Calculate the de Broglie wavelength for a car of mass 500 kg travelling at 20 m s^{-1}. Use your answer to explain why cars don't exhibit wave-like behaviour.

◗ Remember: momentum = mass × velocity.

Module B: end-of-module questions

Speed of light in free space $c = 3 \times 10^8$ m s^{-1}

Planck constant $h = 6.63 \times 10^{-34}$ J s

Electron charge $= 1.6 \times 10^{-19}$ C

1 A 12 Ω resistor is connected in a circuit with a 3 V battery of negligible internal resistance.

   a Draw a circuit diagram to show this circuit. Add arrows to indicate the directions of conventional current flow, and of electron flow.

   b Calculate the current which flows in the circuit. How much charge flows through the resistor each second?

   c What is the potential difference across the resistor?

   d Calculate the energy transferred in the resistor each second.

2 a Write down an equation linking charge, energy and potential difference. Explain how this equation is used to define potential difference.

   b Use your answer to part **a** to define the volt.

   c The electromotive force (e.m.f.) of a cell can be defined as 'the energy transferred per unit charge when the cell pushes 1 coulomb of charge around a circuit'. Explain how this definition is related to the equation you have stated in part **a**.

3 In the diagram, a resistor R and a thermistor T are connected to a cell of negligible internal resistance.

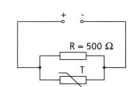

   a Are the two resistors R and T connected in series or in parallel with one another?

   b A current of 10 mA flows through the resistor R. Calculate the p.d. across it.

   c A current of 4 mA flows through T. What is its resistance?

   d The thermistor T is heated so that its resistance decreases. Will the current through it increase, decrease or stay the same?

   e Will the current through R increase, decrease or stay the same?

4 In an experiment to investigate the resistance of an alloy of copper, some students use a 0.56 m length of copper alloy wire of diameter 0.4 mm.

a Calculate the resistance of the wire at room temperature. (Resistivity of the copper alloy at room temperature = 3.4×10^{-7} Ω m.)

The students find that, when a p.d. of 6.0 V is applied across the wire, a current of 400 mA flows through it. Doubling the p.d. to 12.0 V results in a current of 690 mA. They notice that the wire is now hot.

b Sketch a current–voltage characteristic graph for the copper alloy wire, to show the results that you would expect the students to obtain if they measured current and voltage over a greater range of values. State and explain whether the copper alloy wire is an ohmic conductor.

5 A battery of e.m.f. 12.0 V and internal resistance 2.0 Ω is connected as shown to two 5.0 Ω resistors.

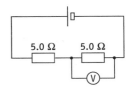

a Calculate the potential difference measured by the high-impedance voltmeter V.

b Calculate the terminal p.d. across the battery.

c Explain why the battery's terminal p.d. would increase if the two resistors were replaced with resistors each of value 5 kΩ.

6 In normal use, a 150 W lamp is found to draw a current of 1.5 A from a supply.

a Calculate the resistance of the lamp when it is in use.

b If the lamp is left switched on for 24 hours, how many kWh of energy are transferred?

7 The diagram shows an arrangement to investigate the force on a current-carrying conductor in a magnetic field. The two magnets provide a uniform magnetic field of flux density 0.15 T. The length of the wire in the field is 4 cm.

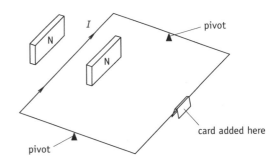

a Calculate the force on the wire due to the magnetic field when a current of 2.4 A flows through it.

b As shown in the diagram, the wire is balanced. A small piece of card of weight 6×10^{-3} N is hung from the wire. This unbalances the wire. What current is now needed to restore the balance?

c Describe briefly how this arrangement could be used to determine the value of an unknown electric current.

8 a A student wrote 'All members of the electromagnetic spectrum travel at the same speed, approximately 3×10^8 m s^{-1}.' Explain how this statement must be modified to make it correct.

b An atom emits a photon of energy 4×10^{-19} J. Calculate the photon's wavelength.

c State which region of the electromagnetic spectrum the photon belongs to.

9 In an experiment to measure the work function of potassium, monochromatic ultraviolet radiation of wavelength 300 nm is shone on the metallic surface of some potassium. Electrons are emitted by the potassium. The fastest-moving electrons are found to have kinetic energy of 2.2 eV.

a Explain what is meant by the term *monochromatic*.

b Calculate the energy of a single photon of the ultraviolet radiation, in eV.

c Explain what is meant by the term *work function*.

d Calculate the work function of potassium, in eV.

e Most of the electrons emitted by the potassium have a kinetic energy less than 2.2 eV. Explain why this is so.

10 a Electrons may be said to exhibit wave-like behaviour. Describe briefly one way in which this behaviour may be demonstrated.

b Atoms may also show wave-like behaviour. Calculate the de Broglie wavelength of an atom whose momentum is 3×10^{-20} kg m s^{-1}.

Module C: Wave properties

Module B looked at the problem of wave–particle duality. The behaviour of light can be explained in terms of two different models: particles (photons) and waves. The same is true for things such as electrons which we normally think of as particles. This module explores in detail the behaviour of waves, particularly in relation to light and other electromagnetic radiation; sound; and vibrating strings.

There are three blocks in this module.

- **Block C1** shows how the reflection and refraction of light can be represented by rays. Snell's law shows how to predict where a refracted ray will go.

- **Block C2** builds up a general description of waves.

- **Block C3** shows how this understanding can be used to explain interference and diffraction.

The reflection and refraction of light can be explained in terms of wave or particle models. However, interference, diffraction and polarisation are wave properties which cannot be explained in terms of particles.

Block C1: Rays, pages 58–61

Ideas from GCSE	Content outline of Block C1
<ul><li>Light can be reflected and refracted</li><li>Total internal reflection and its use in optic fibres</li></ul>	<ul><li>Representing light using rays</li><li>Laws of reflection</li><li>Why refraction happens</li><li>Laws of refraction, including Snell's law</li><li>Definition of refractive index n</li><li>Total internal reflection and the critical angle C</li><li>Applications of TIR in optic fibres</li></ul>

Block C2: Wave definitions, pages 62–67

Ideas from GCSE	Content outline of Block C2
<ul><li>Waves transfer energy without transferring matter</li><li>Longitudinal and transverse waves</li><li>Frequency, wavelength, amplitude and speed of a wave</li></ul>	<ul><li>Longitudinal and transverse waves</li><li>Polarisation of transverse waves</li><li>Wave quantities: displacement, amplitude, wavelength, period, frequency, phase difference</li><li>Energy transfer by a wave</li><li>Intensity, spreading out and absorption</li></ul>

Block C3: Superposition effects, pages 68–73

Ideas from GCSE	Content outline of Block C3
<ul><li>Reflection, refraction and diffraction of waves</li></ul>	<ul><li>Constructive and destructive interference</li><li>Observing interference patterns for light, sound and microwaves</li><li>Determining the wavelength of light</li><li>Principle of superposition</li><li>Standing waves on strings, in air columns and for microwaves</li></ul>

End-of-module questions, pages 74–75

Reflection and refraction

To explain how light behaves, we can think of light travelling as *rays*. A ray travels in a straight line. It will change direction if:

- it is *reflected* (when it strikes a surface);
- it is *refracted* (when it passes from one material to another).

▶▶ *We can also think of light as waves – see pages 62–63.*

Laws of reflection

The **laws of reflection** tell you where a ray will go when it is reflected. The **normal** is the line at 90° to the reflecting surface at the point where the incident ray strikes it.

- **Law 1:** Angle of incidence = angle of reflection, $i = r$ (angles measured from normal to ray).
- **Law 2:** Incident ray, reflected ray and normal are all in the same plane.

These laws apply not only for flat surfaces but also for curved and rough surfaces.

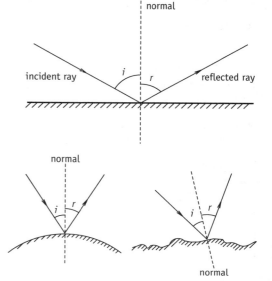

✓ *Quick check 1*

Refraction: when it happens

Light travels fastest in a vacuum. It travels more slowly in other media. When light changes speed (because it travels from one medium to another), it is *refracted*.

- If a ray enters a medium head-on (angle of incidence $i = 0$), it travels straight on.
- If a ray enters a medium obliquely, it bends.

▶▶ *For an explanation of refraction in terms of waves, see page 63.*

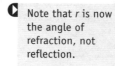

❶ Note that r is now the angle of refraction, not reflection.

Laws of refraction

As with reflection, angles are measured from the normal to the ray.

- **Law 1: Snell's law** (see opposite) explains how the angles of incidence and refraction are related.
- **Law 2:** Incident ray, refracted ray and normal are all in the same plane.

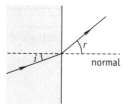

light speeding up: ray bends away from normal

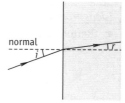

light slowing down: ray bends towards normal

✓ *Quick check 2*

Refractive index *n*

The **refractive index *n*** of a medium relates the speed of light in the medium to the speed of light in free space (a vacuum).

$$\text{refractive index } n = \frac{\text{speed of light in free space}}{\text{speed of light in medium}} = \frac{c_0}{c_{medium}}$$

In a medium of refractive index 2, light travels at half its speed in free space. Some values are worth remembering:

- $n_0 = 1$ (by definition)
- $n_{air} = 1.00$ (to 2 decimal places)
- $n_{water} = 1.33$
- $n_{glass} \sim 1.5$ (depending on the composition of the glass)

✓ *Quick check 3*

Snell's law

For a ray passing from air into a medium of refractive index *n*, the angle of incidence *i* and the angle of refraction *r* are related by:

$$n = \frac{\sin i}{\sin r}$$

Where a ray passes from one medium to another (speeds c_i and c_r, refractive indices n_i and n_r), first calculate the relative refractive index using:

$$n = \frac{c_i}{c_r} = \frac{n_r}{n_i}$$

Worked example

A ray of light travels from glass ($n_i = 1.5$) into water ($n_r = 1.33$) with an angle of incidence *i* of 30°. Calculate the angle of refraction *r*.

Step 1 Calculate the relative refractive index from the values for the two materials:

$$n = n_r \,/\, n_i = 1.33 \,/\, 1.5 = 0.887$$

Step 2 Substitute values into the Snell's law equation, rearrange and solve:

$$n = \sin i \,/\, \sin r$$
$$0.887 = \sin 30° \,/\, \sin r$$
$$\sin r = \sin 30° \,/\, 0.887 = 0.564$$
$$r = 34°$$

The value is less than 1 because the light speeds up as it enters the water.

The refractive index doesn't change much, so the change of direction is small.

✓ *Quick check 4*

❓ Quick check questions

1 What are the values of the angles of incidence and reflection (upper diagram)?

2 Does the ray speed up or slow down when it enters medium 2 (lower diagram)?

3 Which medium in question **2** has the higher refractive index?

4 A ray of light, travelling through air, strikes a glass surface with an angle of incidence of 40°. The refractive index of the glass is 1.47. Draw a diagram to show the situation. Calculate the angle of refraction.

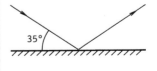

Total internal reflection

Total internal reflection (TIR) of light may occur when a ray is travelling inside a glass block. The ray reaches the edge of the block; what happens next depends on the angle of incidence i.

Note that there is always a weaker reflected ray as well.

- **Total** – because 100% of the light is reflected.
- **Internal** – because the ray is reflected *inside* the material.
- **Reflection** – because the light is reflected, not refracted.

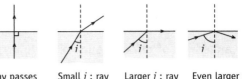

Ray passes straight through ($i = 0°$)

Small i : ray refracted as it leaves block. Some is reflected

Larger i : ray refracted parallel to edge of block. i = critical angle C

Even larger i: ray entirely reflected inside block

Critical angle *C*

Total internal reflection can only happen when a ray travelling through a material of *higher* refractive index reaches the boundary with a material of *lower* refractive index. It occurs for any angle of incidence equal to or greater than the **critical angle *C***.

At the critical angle, $i = C$ and $r = 90°$. Hence $\sin i = \sin C$ and $\sin r = 1$. Snell's law then gives:

$$n = \frac{1}{\sin C} \quad \text{or} \quad \sin C = \frac{1}{n}$$

For glass of refractive index 1.5, $\sin C = 1/1.5 = 0.667$, and $C = 42°$, approximately.   ✓ *Quick check 1, 2*

Using TIR in optic fibres

A ray of light can travel along inside a solid glass fibre. Each time it reaches the outer surface of the glass it is reflected back inside, since i is nearly 90°.

Practical optic fibres

Optic fibres are made from glass or plastic, surrounded by a coating of material with a slightly lower refractive index.

Rays which travel straight down the centre of the fibre have the shortest route and take least time. Oblique rays have further to travel, and take longer.

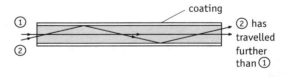

coating

①

② has travelled further than ①

✓ *Quick check 3*

Transmitting data

Optic fibres carry data in digital form. A ray of light from a laser is modulated (switched on and off) at high frequency to encode the data, rather like Morse code. Problems arise if rays can travel along different paths inside the fibre.

a single pulse enters the fibre

the pulse is 'smeared': some rays have taken longer than others

This smearing of a pulse is called **multipath dispersion**. To avoid this problem, most fibres are made with a very narrow core so that all rays pass virtually straight down the middle.

✓ *Quick check 4*

Advantages of optic fibres

Optic fibres have made possible the Internet. They are used for:

- telecommunications networks (carrying telephone messages),
- cable television,
- links between computers (for high-speed data transfers).

Digital signals are less susceptible to noise than analogue signals. Because of the high frequency of light, optic fibres can carry vastly more data than a current in a cable of comparable size. They are very difficult to bug. Because the light is only weakly absorbed, signals can travel many kilometres before they become so weak that they need to be regenerated.

? Quick check questions

1 Calculate the critical angle for glass of refractive index 1.6.

2 Calculate the critical angle at the interface between glass ($n = 1.6$) and water ($n = 1.33$).

3 Explain why the central core of an optic fibre must be coated with a material of *lower* refractive index.

4 Different wavelengths of light travel at different speeds through glass. Explain why white light could not be used for long-distance information transfer. Why is laser light suitable?

Wave representations

We see waves on the surface of water. They travel across the surface of the water, transferring energy; the molecules of the water move up and down. A wave is a periodic disturbance of the water.

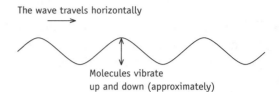

The wave travels horizontally

Molecules vibrate
up and down (approximately)

The top diagram represents the wave as an idealised **sine wave**. This idea can be used as a model for other phenomena:

- **Sound waves** travel through air (or any other medium). The particles of the medium vibrate back and forth as the wave travels along, as shown in the lower diagram.

- **Light** (and other **electromagnetic waves**) do not require a medium. They are a periodic disturbance of the electric and magnetic fields through which they are travelling. These fields vary at right angles to the direction of travel of the wave.

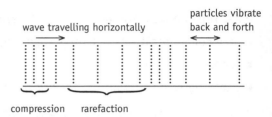

wave travelling horizontally

particles vibrate
back and forth

compression   rarefaction

Although we may represent a sound wave using a sine curve, the particles move back-and-forth, not up-and-down.

Transverse and longitudinal waves

Transverse waves can be made to travel along a stretched rope, by moving one end up and down (or from side to side). Both types of wave can be demonstrated using a long spring; for longitudinal waves, the end of the spring must be pushed back and forth. However, it is simplest to represent waves, transverse or longitudinal, as sine waves.

✓ *Quick check 1, 2*

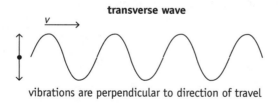

transverse wave

v

vibrations are perpendicular to direction of travel

longitudinal wave

v

vibrations are back-and-forth, along direction of travel

Polarisation

Light (and other transverse waves) can be **polarised**.
In unpolarised light, the electric and magnetic fields vibrate in all directions perpendicular to the direction of travel. After passing through a piece of Polaroid, each vibrates in only one direction.

Only transverse waves can be polarised. If a wave can be polarised, it must be transverse – that's how we know electromagnetic waves are transverse.

B   E   v   B-field   E-field

magnetic (B) and electric (E) fields vibrate perpendicular
to one another, and to the direction of travel

Wave fronts and rays

The *ripple tank* shows another way to represent waves. We draw wave fronts, as though we were looking down on the ripples from above.

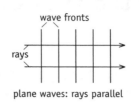

plane waves: rays parallel

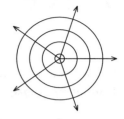

circular waves spreading out from a point source

We can add rays; these are always perpendicular to the wave front.

Note that the separation of the wave fronts is constant.

Reflection and refraction

All waves can be reflected and refracted.

Notice that the ray is always at 90° to the wave fronts.

▶▶ *You should be able to relate these diagrams to the corresponding ray diagrams on pages 58–59.*

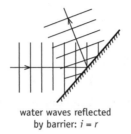

water waves reflected by barrier: $i = r$

slower moving in medium ②
wave front lags behind

medium ① medium ② shallower water

waves travel more slowly in medium ②
and are refracted

submerged barrier

✓ *Quick check 3, 4*

Changing wavelength

When a wave enters a medium where it travels more slowly, its wavelength decreases. Its frequency remains constant.

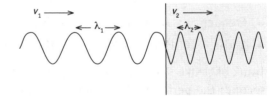

? ## Quick check questions

1 Classify as transverse or longitudinal: light, sound, water, infrared waves.

2 A guitarist plucks a string. A wave travels along the string. Is this longitudinal or transverse?

3 Draw a ray diagram to show a single ray being reflected by a mirror placed at 45° to its path. Add wave fronts to show how these are reflected by the mirror.

4 Copy and complete the diagram to show what happens when waves enter a medium where they travel more slowly. The boundary is parallel to the wave fronts.

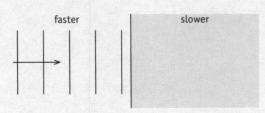

faster   slower

Wave quantities

Several quantities are needed to fully describe a wave: **amplitude**, **wavelength**, **frequency**, **phase**. Learn how they are related; take care not to confuse them.

Wavelength and amplitude

- The **displacement *y*** is the distance moved by any particle from its undisturbed position.
- The **wavelength** λ of a wave is the distance between adjacent crests (or troughs), or between any two adjacent points which are at the same point in the cycle (i.e. which are **in phase** with each other).
- The **amplitude *A*** of a wave is the maximum displacement of any particle.

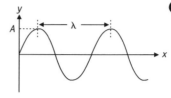

Horizontal axis = distance

❶ Amplitude is the height of a crest measured from the horizontal axis, not from crest to trough.

✓ *Quick check 1*

Period and frequency

- The **period *T*** is the time for one complete cycle of the wave.
- This is related to the wave's **frequency *f***: $T = 1/f$ (or $f = 1/T$).
- Frequency is measured in **hertz** (Hz). 1 Hz = 1 wave/s = 1 s^{-1}.
- 1 kHz = 10^3 Hz   1 MHz = 10^6 Hz   1 GHz = 10^9 Hz

Think of it like this: the frequency is the number of waves per second; the period is the number of seconds per wave.

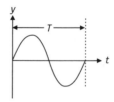

❶ Take care! The horizontal axis of this graph is *time*, not distance.

✓ *Quick check 2*

Phase difference

Two waves may have the same wavelength but be out of phase (out of step) with one another. Phase difference is expressed as a fraction of a cycle, or in **radians** (rad) or **degrees** (°).

- 1 cycle = 1 complete wave = 2π rad = 360°
- $\frac{1}{2}$ cycle = π rad = 180° $\frac{1}{4}$ cycle = $\pi/2$ rad = 90°

two waves in phase

phase difference=½ cycle

phase difference=¼ cycle

✓ *Quick check 3*

Measuring frequency

To find the frequency of a sound wave, plug a microphone into an oscilloscope (c.r.o.) and use it to display the sound.

- **Step 1** Adjust the timebase setting to give two or three complete waves on the screen.

  Timebase setting = 0.02 s div^{-1}.

- **Step 2** Measure the width of a number of complete waves.

  Two waves occupy 5.0 divisions.

- **Step 3** Calculate the time represented by this number of divisions.

  Time = 5.0 div × 0.02 s div^{-1} = 0.10 s.

- **Step 4** Calculate the frequency = number of waves / time.

  Frequency = 2 waves / 0.10 s = 20 Hz.

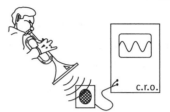

> ⟨ Check that the 'variable timebase' knob is in the 'calibrated' position.

> ⟨ Timebase setting may be given in divisions or centimetres.

✓ *Quick check 4*

? ## Quick check questions

1 What quantities are represented by p and q in the diagram? What are their values?

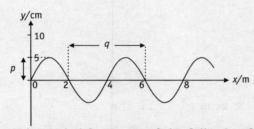

2 Calculate the period for waves of the following frequencies: 2 Hz, 2 kHz, 0.5 MHz.

3 On the same axes, sketch two waves with a phase difference of π radians; one wave has twice the amplitude of the other.

4 An oscilloscope is set with its timebase at 5 ms cm^{-1}. An alternating signal gives four complete waves across the 6 cm screen. What is the frequency of the signal?

Wave speed

Waves are one way in which energy is transferred from place to place. How quickly they do this depends on their speed, which may be anything up to c, the speed of light in free space, 3×10^8 m s^{-1}.

Speed, frequency and wavelength

The waves we have considered so far are described as **progressive waves**. They travel through space. The **speed v** of the wave tells us how fast it moves. The speed is the distance travelled per second by a crest.

▶▶ *The opposite of a progressive wave is a standing or stationary wave – see page 72.*

Speed v is related to frequency f and wavelength λ by

$$\textbf{speed = frequency} \times \textbf{wavelength} \qquad v = f\lambda$$

If a 'train' of f waves, each of length λ, passes a point in 1 s, the total length of the train is $f\lambda$. This is the length of the waves passing per second, i.e. the speed of the wave.

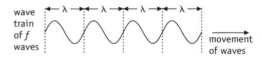

A note on units

- Frequency f is in hertz (Hz).
- Wavelength λ is in metres (m).

Since 1 Hz = 1 s^{-1}, multiplying $f \times \lambda$ gives a result in m Hz, or m s^{-1}. This is the correct unit for speed.

Worked examples

1 An observer, standing at the end of a pier, observes one wave passing by every 8 s. The distance between adjacent peaks is 12 m. Calculate the speed of the waves.

Step 1 Calculate the frequency of the waves.

$$f = \frac{1}{8\,\text{s}} = \textbf{0.125 Hz}$$

Step 2 Note down the wavelength of the waves.

$$\lambda = \textbf{12 m}$$

Step 3 Calculate the wave speed.

$$\textbf{speed } v = f\lambda = \textbf{0.125 Hz} \times \textbf{12 m = 1.5 m s}^{-1}$$

2 Calculate the wavelength of an electromagnetic wave (speed = 3×10^8 m s^{-1}) of frequency 100 GHz.

Step 1 Write down quantities; convert to scientific notation (powers of 10).

$$v = 3 \times 10^8 \text{ m s}^{-1}, f = 100 \text{ GHz} = 100 \times 10^9 \text{ Hz}, \lambda = ?$$

Step 2 Rearrange the equation, substitute values and solve.

$$\lambda = \frac{v}{f} = \frac{3 \times 10^8 \text{ m s}^{-1}}{100 \times 10^9 \text{ Hz}} = 3 \times 10^{-3} \text{ m}$$

Note that it is simplest to change units such as GHz to powers of 10; then enter them into your calculator in this form. You could give this answer as $\lambda = 3$ mm.

✓ *Quick check 1, 2*

Quick check questions

1 Calculate the speed of ripples whose wavelength is 3 mm and whose frequency is 15 Hz.

2 Calculate the frequency of a sound wave if its wavelength in air is 11 mm. (Speed of sound in air = 330 m s^{-1}.)

Interference and diffraction

What happens when two waves meet? Two snooker balls would bounce off one another, but waves behave differently. They show behaviour known as **interference**.

Constructive interference

two waves arriving
in phase (in step)

resultant is wave of
twice the amplitude

Destructive interference

two waves arriving
out of phase (out of step)

cancel each
other out

✓ *Quick check 1*

Interference of sound

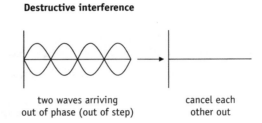

Walking around in the space beyond the two loudspeakers, you can hear points where the sound is loud, and points where it is much softer. These loud and soft points have a regular pattern.

Your ear receives waves from both speakers. Suppose the wavelength of the sound waves is 1 m. If your ear is 4 m from one speaker and 5 m from the other, there is a **path difference** of 1 m for the two waves. They will be *in phase*; they will interfere constructively and you hear a loud sound.

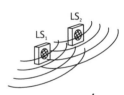

If your ear is 4 m from one speaker and 5.5 m from the other, the path difference is 1.5 m. The waves will be *out of phase*; they will interfere destructively and you hear no sound (or a very faint sound).

- For constructive interference, path difference = $n\lambda$.
- For destructive interference, path difference = $(n + \frac{1}{2})\lambda$.

Interference of other waves

The same effect can be shown for:

- *ripples* – use two dippers attached to a vibrating bar in a ripple tank;
- *microwaves* – direct the microwaves through two gaps in a metal plate;
- *light* – the 'Young's slits' experiment – see page 70.

Diffraction of ripples

When ripples pass through a gap, they spread out into the space beyond. The effect, which is known as **diffraction**, is greatest when the width of the gap x is similar to the wavelength of the ripples λ.

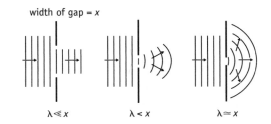

width of gap = x

$\lambda \ll x$ $\lambda < x$ $\lambda \approx x$

✓ *Quick check 2*

Explaining diffraction

When light from a laser is shone through a single slit, a **diffraction pattern** of light and dark interference bands (called '*fringes*') is seen on the screen. We picture waves spreading out from all points in the slit. Each point on the screen receives waves from each point in the slit. These waves interfere.

- Where all the interfering waves cancel each other out, we see a dark fringe (*destructive* interference).
- Where all the interfering waves add up, we see a bright fringe (*constructive* interference).

Coherent sources

To observe an interference pattern where two sets of waves overlap, they must be **coherent**. This means they must have the same wavelength and frequency; also, the phase difference between them must be constant.

The two loudspeakers in the diagram opposite are coherent sources. They are connected to the same signal generator, so they vibrate back and forth in step with each other.

Light from a lamp is not usually coherent. It is emitted as photons, and they do not keep in step with one another. Laser light is coherent; its photons remain in step between source and screen.

✓ *Quick check 3*

❓ Quick check questions

1 What will be observed if two waves, in phase and one having twice the amplitude of the other, interfere?

2 Draw a ripple diagram (like those at the top of this page) to show ripples of wavelength λ being diffracted by a gap of width 2λ. Draw a second diagram to show what happens if ripples of twice this wavelength pass through the same gap.

3 Two dippers are used to produce an interference pattern in a ripple tank. Are they a pair of coherent sources?

Young's slits experiment

Light shows *interference*. To produce two rays, light is shone through a pair of parallel slits. Where the light falls on a screen beyond the slits, light and dark interference 'fringes' are seen.

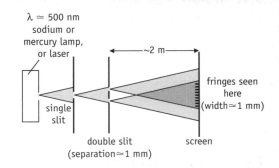

- The single slit acts as a narrow source of light, shining on the double slit. Alternatively, a laser can be shone directly on the double slit.

- As light passes through each slit, it spreads out into the space beyond. This is *diffraction* – see page 69.

- The fringe separation can be measured using a travelling microscope. Increasing the slit–screen distance makes the fringes wider but dimmer.

✓ *Quick check 1*

Explaining the interference fringes

Each point on the screen receives light waves from both slits (S_1 and S_2).

- **A** is the point on the screen directly opposite the point midway between the two slits. If waves leave the two slits in phase with one another, they will arrive at point **A** in phase. They will interfere constructively and a bright fringe will be seen. Path difference = 0.

- **B** is the centre of the first dark fringe. Waves from S_1 have a shorter distance to travel than light from S_2. The two waves arrive out of phase and interfere destructively. Path difference = $S_2\textbf{B} - S_1\textbf{B} = \lambda/2$.

- **C** is the centre of the next bright fringe. The two waves arrive in phase, but one has travelled further than the other. Path difference = λ.

Therefore:

- A bright fringe is seen where the two waves arrive in phase; path difference = $n\lambda$.

- A dark fringe is seen where they arrive out of phase; path difference = $\left(n + \frac{1}{2}\right)\lambda$.

- In-between positions have in-between path differences which give rise to intermediate brightnesses.

✓ *Quick check 2*

Measuring the wavelength of light

The Young's slits experiment provides a method for determining λ, which is related to the screen distance D, slit separation a and fringe width x by:

$$\lambda = ax/D$$

Note that, for white light, this can only give an average value of λ since many wavelengths are present. Laser light is monochromatic (a single wavelength) so the fringes are clearer and a more accurate value of λ can be found.

You may find it easier to remember the formula as $\lambda D = ax$: largest quantity (D) times smallest (λ) equals the other two multiplied together.

✓ *Quick check 3*

Worked example

Laser light of wavelength 648 nm falls on a pair of slits separated by 1.5 mm. What will be the separation of the interference fringes seen on a screen 4.5 m from the slits?

Step 1 Write down what you know, and what you want to know:

$$\lambda = 648 \text{ nm}, a = 1.5 \text{ mm}, D = 4.5 \text{ m}, x = ?$$

Step 2 Rearrange the equation, substitute values and solve:

$$x = \frac{\lambda D}{a} = \frac{648 \times 10^{-9} \text{ m} \times 4.5 \text{m}}{1.5 \times 10^{-3} \text{ m}} = 1.9 \times 10^{-3} \text{ m}$$

So the fringe width seen on the screen will be 1.9 mm.

✓ Quick check 4

Microwaves, ripples

Arrange metal plates to form two vertical slits (gaps), approximately 5 cm wide and separated by 5 cm. Direct microwaves at the slits from a single source; a pattern of regularly spaced high and low intensity 'fringes' are detected in the space beyond.

The same effect can be seen using ripples in a ripple tank. Pass the ripples through a barrier with two gaps; regularly spaced 'fringes' are detected beyond the barrier.

? Quick check questions

1 Give the symbol and approximate size for each of the following in the Young's slits experiment: slit–screen distance; slit separation; fringe separation; wavelength of light.

2 What can you say about the path difference between two waves which show destructive interference?

3 If the slit separation a is doubled, how will the fringe width x be changed?

4 White light is directed onto a pair of slits separated by 1.0 mm. Interference fringes are observed on a screen at a distance of 1.8 m. Five fringes have a width of 5.0 mm. Estimate the wavelength of the light. Why is your answer an estimate?

Superposition and standing waves

When two or more waves meet, the result is found by the **principle of superposition**. At any instant, the resultant displacement is simply the sum of the displacements of the individual waves. Constructive and destructive interference are obvious examples of this idea. It also explains the formation of **standing waves**.

Standing waves on a stretched string

The vibrator sends waves along the string. They reflect at the other end. The outgoing and reflected waves then interfere. At certain frequencies, a standing wave (or **stationary wave**) pattern of loops is formed.

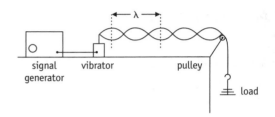

- At certain points – **nodes** – the two waves interfere destructively. There is no vibration. There are nodes at the ends of the string.

- Half-way between the nodes are **antinodes**. The string vibrates with a large amplitude. When the vibration has its maximum amplitude, the two waves are interfering constructively.

- Changing the frequency slightly causes the standing wave to disappear. Changing the length, tension or thickness of the string causes the standing waves to appear at different frequencies.

- The wavelength of the wave is *twice* the distance from one node to the next.

✓ *Quick check 1*

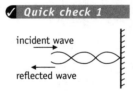

Conditions for a standing wave

Two identical but oppositely travelling waves interfere with each other to form a standing wave. Often, one wave is a reflection of the other. For example, when microwaves are reflected by a metal plate, a standing wave pattern is formed.

✓ *Quick check 2*

Using the principle of superposition

The diagrams show the two waves which make a standing wave. They are shown at three instants in time. You can see that the waves are progressive waves, travelling in opposite directions.

Below these are the resultant waves. These are worked out by adding the displacements of the two progressive waves.

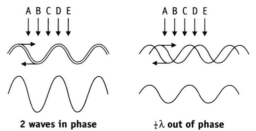

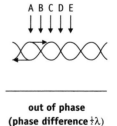

- Points A, C and E are nodes; the two waves always cancel here.

- Points B and D are antinodes; the displacement here varies up and down.

✓ *Quick check 3*

Air columns

When the frequency of the loudspeaker is changed, a point is reached where the note becomes much louder. Sound waves are reflected by the water and a standing wave has formed in the air column inside the cylinder. There is a node at the foot of the air column and a node at the top.

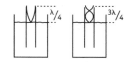

At the lowest frequency at which this occurs, the length of the air column is one quarter of the wavelength of the sound. A standing wave is formed again at three times this frequency, with three-quarters of a wave fitting in the column.

✓ *Quick check 4*

Microwaves

Direct microwaves at a vertical metal plate. Reflected waves interfere with incoming waves to form a standing wave pattern. Detect nodes (zero intensity) and antinodes (high intensity) between source and plate.

Quick check questions

1 A string of length 1.2 m is stretched and vibrated so that a standing wave consisting of two loops is formed. Sketch this, and calculate the wavelength of the waves on the string.

2 Microwaves are directed at a sheet of steel. A detector is used to investigate the intensity of the waves between the source and the plate. A pattern of high and low intensity regions is found; the separation of adjacent high intensity regions is 1.5 cm. What is the wavelength of the microwaves?

3 Explain why there is a node at point C in the diagram at the foot of the opposite page.

4 In a vibrating air column experiment, the air column is 20 cm long. The lowest frequency which produces a standing wave is 400 Hz. Calculate the wavelength and speed of the sound waves.

Module C: end-of-module questions

Speed of light in free space = 2.998×10^8 m s^{-1}

1 State the two laws of reflection of light. Illustrate your answer with a diagram.

2 A ray of light travels through glass at a speed of 2.043×10^8 m s^{-1}. It strikes the surface of the glass at an angle of incidence of 20° and is refracted as it passes out into the vacuum beyond.

   a Calculate the refractive index of the glass.

   b Calculate the critical angle at the glass–vacuum interface.

   c Calculate the angle of refraction of the ray as it leaves the glass.

3 The diagram shows two rays of light travelling along a narrow fibre made of glass of refractive index 1.51. Ray A travels along the central axis of the fibre. Ray B reflects back and forth along the fibre. Its angle of incidence with the inner surface of the fibre is 85°.

   a Calculate the speed of light in the glass.

   b Explain why ray B is reflected each time it strikes the inner surface of the glass.

   c The fibre is 50 km long. In travelling from one end to the other, ray B travels further than ray A.

   (i) Calculate the extra distance travelled by ray B as it travels the length of the fibre.

   (ii) Calculate the extra time taken by ray B.

   d Explain why this is a problem when transmitting digital signals along an optic fibre.

   e Explain how this problem can be overcome by the use of fibres with very narrow cores (monomode fibres).

4 Sound waves from a vibrating transducer are transmitted along a steel rail. Their speed in the steel is 6000 m s^{-1}.

   a The frequency of the vibrations is 50 kHz. Calculate their wavelength in the steel.

   b If the frequency of the vibrations is decreased, how will their wavelength be affected?

5 You are provided with a microwave generator, two metal plates, and a microwave detector. Describe how you would demonstrate the diffraction of microwaves using this equipment.

6 Laser light of wavelength 648 nm is passed through a pair of parallel slits; on a screen 5.7 m away, a pattern of light and dark fringes is seen.

   a Using the terms *constructive interference* and *path difference*, explain how a bright fringe is formed.

   b The width of 10 of the fringes is found to be 2.8 cm. Calculate the separation of the two slits.

7 Here are three types of wave:

   light waves   microwaves   sound waves

   a Which of these are longitudinal waves?

   b Which of these can be polarised?

8 Stationary waves can be created by plucking a stretched string.

   a With the aid of a diagram, explain what is meant by a node and an antinode.

   b Explain how such a standing wave is formed when the string is plucked.

9 A string is stretched horizontally. One end is moved up and down at a frequency of 30 Hz, so that a train of waves travels along the string. The wavelength of these waves is found to be 4 cm.

   a Are these waves longitudinal or transverse? Explain your answer.

   b Calculate the velocity of the waves.

   c When the waves reflect from the other end of the string, a standing wave pattern is formed. What is the separation of adjacent nodes in this pattern?

Appendix 1: Accuracy and errors

Physicists try to make their observations as accurate as possible. Errors in measurements arise in a number of ways and, as an experimentalist, you should try to minimise errors.

Systematic errors

These can arise in a number of ways:

- **Zero error**: e.g. an ammeter does not read zero when no current is flowing through it. If it reads +0.05 A, all of its readings will be too high. Either correct the meter to read zero, or adjust all readings to take account of the error.

- **Incorrect calibration** of an instrument: e.g. an ammeter that reads zero when no current flows, but all other readings are consistently too low or too high. It may read 9.9 A when 10.0 A is flowing. Again, either correct the meter, or adjust all readings.

- **Incorrect use** of an instrument: e.g. screwing a micrometer too tightly, or viewing a meniscus from an angle. Learn the correct technique for using instruments and apparatus.

- **Human reaction**: e.g. when starting and stopping a stopclock. You may always press the button a fraction of a second after the event.

Systematic errors can be reduced or even eliminated. This increases the **accuracy** of the final result.

Random errors

These often arise as a result of judgements made by the experimenter:

- **Reading from a scale.** You may have to judge where a meter needle is on a scale – what is the nearest scale mark? What fraction of a division is nearest to the needle?

- **Timing a moving object.** When did it start to move? When did it pass the finishing line? You have to judge.

The conditions under which the measurement is made can vary:

- **Equipment** can vary. One trolley may have more friction than another. Two apparently identical resistors may have slightly different values.

- **Samples of materials** may be different. Two lengths of wire from the same reel may have slightly different compositions.

- **Conditions** can vary. Room temperature may change and affect your results.

Some measurements are intrinsically random:

- **Radioactive decay.** If you measure the background radiation in the laboratory for 30 s, you are likely to find slightly different values each time.

Random errors can be reduced, but it is usually impossible to eliminate them entirely. Reducing random errors increases the **precision** of the final result.

Reducing random errors

Here are some ways to reduce random errors.

- **Make multiple measurements**, and find the mean (average). Roughly speaking, taking four measurements reduces the error by half; 100 measurements will divide the error by 10.
- **Plot a graph**, and draw a smooth curve or a straight line through the points.
- **Choose a suitable instrument** to reduce errors of judgement, e.g. using light gates and an electronic timer instead of timing with a stopwatch. You need to think critically about the instrument: does it introduce other sources of error?

Expressing errors

Here are two ways in which the error or uncertainty in a final result can be expressed.

- **Use significant figures**: a calculation may give $R = 127\ \Omega$. If the errors are small, you may wish to quote this as $130\ \Omega$; if the errors are large, as $100\ \Omega$.
- **Use ± errors**: by considering the errors in individual measurements, you may be able to show the degree of uncertainty in the above result. Small error: $R = (127 \pm 2)\ \Omega$; larger error: $R = (130 \pm 10)\ \Omega$.

Summary

- Think critically about the equipment and methods you use.
- Reduce random errors to increase the precision of your results.
- Reduce systematic errors to increase the accuracy of your results.
- Indicate the extent of error or uncertainty in individual results, and in the final result.

Appendix 2: Data and formulae for question papers

In end-of-module question papers, you will be supplied with a long list of data and formulae. **Part 1** below shows the data and formulae relevant to modules A–C.

You are expected to recall many formulae. **Part 2** below shows those relevant to modules A–C.

Part 1: Data and formulae supplied in question papers

acceleration of free fall, $g = \textbf{9.81 m s}^{-2}$

speed of light in free space, $c = \textbf{3.00} \times \textbf{10}^{8} \textbf{ m s}^{-1}$

elementary charge, $e = \textbf{1.60} \times \textbf{10}^{-19} \textbf{ C}$

the Planck constant, $h = \textbf{6.63} \times \textbf{10}^{-34} \textbf{ J s}$

rest mass of electron, $m_{e} = \textbf{9.11} \times \textbf{10}^{-31} \textbf{ kg}$

rest mass of proton, $m_{p} = \textbf{1.67} \times \textbf{10}^{-27} \textbf{ kg}$

uniformly accelerated motion, $s = ut + \frac{1}{2} at^{2}$

$$v^{2} = u^{2} + 2as$$

refractive index, $n = 1/\sin C$

It is a good idea to become familiar with these lists of data and formulae, so that you know what is provided on the question papers.

Part 2: Formulae not supplied in question papers

Module A: Forces and motion

speed, $v = s/t$

acceleration, $a = (v - u)/t$

force, $F = ma$

weight, $W = mg$

density, $\rho = m/V$

moment of a force, $T = Fx$

torque of a couple, $T = Fx$

pressure, $p = F/A$

work done, $W = Fd$

kinetic energy, $E_{k} = \frac{1}{2} mv^{2}$

gravitational potential energy, $\Delta E_{p} = mg\Delta h$

power, $P = W/t$

$P = Fv$

stress, $\sigma = F/A$

strain, $\varepsilon = \Delta l/l$

the Young modulus, $E =$ **stress/strain** $= \sigma/\varepsilon$

Module B: Electrons and photons

electric current, $I = \Delta Q/\Delta t$

potential difference, $V = W/Q$, $V = P/I$

electrical resistance, $R = V/I$

resistivity, $\rho = RA/l$

power, $P = I^2R$

$P = V^2/R$

electrical energy, $W = VIt$

resistors in series, $R = R_1 + R_2 + ...$

resistors in parallel, $1/R = 1/R_1 + 1/R_2 + ...$

force on a current-carrying conductor, $F = BIl \sin \theta$

photon energy, $E = hf$

photo-electric effect, $hf = \phi + E_{k\,max} = \phi + \frac{1}{2}mv^2_{max}$

de Broglie equation, $\lambda = h/p = h/mv$

Module C: Wave properties

refractive index, $n = c_i/c_r = n_r/n_i$

$n = \sin i/\sin r$

wave speed, $v = f\lambda$

double-slit interference, $\lambda = ax/D$

Appendix 3: Useful definitions

Here are brief statements of the definitions you need to know. They have been grouped together in clusters of related terms, to help you learn them.

Module A: Forces and motion

scalar quantity	A quantity which has only magnitude.
vector quantity	A quantity which has both magnitude and direction.
displacement	Distance moved in a particular direction.
speed	The rate of change of distance moved with respect to time.
velocity	The rate of change of displacement with respect to time.
acceleration	The rate of change of an object's velocity.
weight	The force caused by a gravitational field acting on an object's mass.
newton (N)	The SI unit of force; 1 N will give a mass of 1 kg an acceleration of 1 m s^{-2}.
work done by a force	The energy transferred when a force moves through a distance; work done = force × distance moved in the direction of the force.
power	The rate at which energy is transferred.
watt (W)	The SI unit of power; 1 watt is 1 joule per second.
moment of a force about a point	The magnitude of a force, times the distance from the point to the line of the force.
torque of a couple	One of the forces times the perpendicular distance between them.
density	For a material: the mass per unit volume.
pressure	The force per unit area on a surface, acting normally to the surface.
Hooke's law	The extension produced in an object is proportional to the load producing it, provided the elastic limit is not exceeded.
elastic limit	The point beyond which an object will not return to its original dimensions.
strain	The extension per unit length produced when an object is stretched or squashed.
stress	The load acting on an object per unit cross-sectional area.
Young modulus	The ratio of stress to strain in a material when it is stretched, provided Hooke's law is obeyed.

Module B: Electrons and photons

coulomb (C)	1 C of charge passes a point when a current of 1 A flows for 1 s.
potential difference (p.d.)	The p.d. between two points is the energy transferred for each 1 C of charge which moves between the points.
volt (V)	1 volt is one joule per coulomb.
electronvolt (eV)	An energy unit; the energy transferred when an electron moves through a p.d. of 1 V.
resistance	The resistance between two points is the ratio of the p.d. between them to the current flowing between them.
ohm (Ω)	The resistance between two points is 1 Ω if a p.d. of 1 V between them causes a current of 1 A to flow.
resistivity	The resistance per unit length of a piece of material of cross-sectional area 1 m^2.
magnetic flux density	The flux density of a magnetic field is equal to the force which acts on unit length of a conductor carrying unit current perpendicular to the field.
tesla (T)	1 tesla is 1 newton per ampere-metre.

Module C: Wave properties

refractive index	When light passes from one material to another, the refractive index is the ratio of the speeds of light in the two materials.
diffraction	Waves spreading out when they pass through a gap.
principle of superposition	For two or more waves meeting at a point: the resultant displacement is the sum of the displacements of the individual waves.

Appendix 4: Electrical circuit symbols

You need to be able to recall and use appropriate circuit symbols; you also need to be able to draw and interpret circuit diagrams which include these symbols.

name of device	symbol
junction of conductors (optional dot)	
conductors crossing (no connection)	
cell	
battery of cells	
open terminals	
indicator or light source	
fixed resistor	
potentiometer (voltage divider)	
light-dependent resistor (LDR)	
thermistor	
ammeter	
voltmeter	
semiconductor diode	

Answers to quick check questions

Module A: Forces and motion

Block A1

Velocity and displacement

1 30 km h^{-1}; 8.3 m s^{-1}

2 200 m s^{-1}

3 40 s

4 Direction changes, so velocity is not constant.

5 9 m s^{-1}

Acceleration

1 4 m s^{-2}

2 3 s

3 −0.6 m s^{-2}; 3750 m

The equations of motion – part 1

1 1.25 m s^{-2}; 250 m s^{-1}

2 19.6 m

3 25 m s^{-1}

4 540 m

5 8.1 m s^{-1}

6 9 s; 189 m

The equations of motion – part 2

1 20 m s^{-1}

2 0.5 m s^{-2}; 95 m

3 25 km

Using vectors

1 1077 km; 22° E of N

2 6.9 m s^{-2}

3 6.6 m s^{-1}; 4.6 m s^{-1}

Block A2

Force, mass, acceleration

1 vector

2 36 N

3 2 m s^{-2} upwards

4 scalar

5 A

Gravity and motion

1 44.1 m

2 228 N; 17.1 m

3 weight: vector; mass: scalar

4 Velocity decreases to slower, steady value.

Force, work and power

1 10 kJ; 4 kJ; 6 kJ

2 1 J = 1 N m = 1 kg m s^{-2} × 1 m

3 200 kJ; 1.96 MJ

4 3000 kJ (= 3 MJ)

5 force

Turning effect

1 20 N m; 14.1 N m

2 3 N and 3 N; 15 N m

3 5 N to right

4 600 kg m^{-3}; 2 m^3

Block A3

Deforming solids

1 0.2 m; 250 N m^{-1}

2 0.002; 1.6 mm

3 200 GPa; brittle

4 150 000 N (= 150 kN)

Forces on vehicles

1 50 kJ; 50 kJ

2 2000 N; 40 000 W (= 40 kW)

3 9 m; 37.5 m; 46.5 m

Module B: Electrons and photons

Block B1

Electric current

1 5 A

2 14 A

3 8 mA (= 0.008 A)

4 6000 C

Resistance

1 5 Ω

2 10 000 V (= 10 kV)

3 2.5 V across each

4 10 Ω

5 18 mA

Ohm's law

1 R = 11 Ω approx.

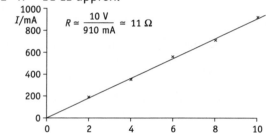

$$R \approx \frac{10 \text{ V}}{910 \text{ mA}} \approx 11 \text{ Ω}$$

2 6 V; 2 Ω

3 0.08 Ω

4 It increases.

p.d. and e.m.f.

1 9 J

2 60 J

3 4 V

4 160 V

5 6 V; 0.2 A; 1 V, 2 V, 3 V; 6 V (= 1 + 2 + 3 V)

6 230 J

Internal resistance and potential dividers

1 3.997 A

2 1.55 V; 19.4 Ω

3 4 V

4

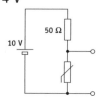

Electrical power

1 450 MJ (= 450 million J, or 450 000 000 J)

2 0.5 W

3 30 W

4 5 A

5 3.6 kWh

Block B2

Magnetic fields

1 repel

2, 3

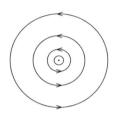

4 field lines further apart

Magnetic forces

1

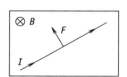

2 no force: WZ and XY; anticlockwise

3 0.05 T

4 the second wire

The ampere and the tesla

1 currents: opposite; forces: repelling

2

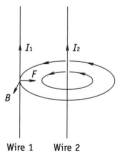

3

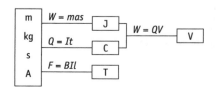

Block B3

The electromagnetic spectrum

1 7×10^{-7} m (or 700×10^{-9} m)

2 radio, infrared, visible, X-rays

3 most energetic: γ-rays; least: radio waves

4 all three

5 500 s

Photons

1 3.315×10^{-19} J

2 10.5 eV

3 2.07 eV

4 654 nm (6.54×10^{-7} m)

5 9.5×10^{-28} kg m s^{-1}

The photoelectric effect

1 an e.m.f.; from − to + (as in any cell)

2 3.62×10^{14} Hz

3 2.9×10^{5} m s^{-1}

Wave–particle duality

1 slower; larger diameter rings

2 particles

3 6.6×10^{-38} m

Module C: Wave properties

Block C1

Reflection and refraction

1 55° (both)

2 slows down

3 medium 2

4 25.9°

Total internal reflection

1 38.7°

2 56.2°

3 otherwise no TIR

4 Signal would be smeared out; laser light is monochromatic (one wavelength).

Block C2

Wave representations

1 Sound is longitudinal, the others transverse.

2 transverse

4 wave fronts closer together

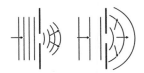

faster   slower

Wave quantities

1 p = amplitude = 5 cm; q = wavelength = 4 m

2 0.5 s; 0.5 ms; 2 μs

3
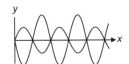

4 133 Hz

Wave speed

1 45 mm s^{-1} (= 0.045 m s^{-1})

2 30 kHz

Block C3

Interference and diffraction

1 single wave of same wavelength, 3 times the amplitude

2
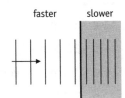

3 yes

Young's slits experiment

1 D; a; x; λ

2 Path difference = (integer + $\frac{1}{2}$) × wavelength.

3 halved

4 555 nm; measurements have error.

Superposition and standing waves

1 $\lambda = 1.2$ m

2 3.0 cm

3 Displacements are always equal and opposite.

4 80 cm; 320 m s^{-1}

Answers to end-of-module questions

Module A: Forces and motion

1 a Velocity has direction, speed has only magnitude.

b vectors: force, velocity, acceleration; scalars: distance, kinetic energy, power

2 a Acceleration = change in velocity / time taken.

b AB: 1.5 m s^{-2}; BC: 0

c
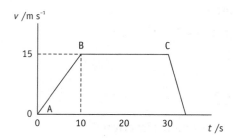

3 a

b Resistive force = forward force = 34.6 kN.

4 a Air resistance (drag) decreases as speed decreases.

b
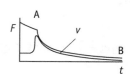

5 a There is a downward pull of 9.8 N on each kg of the mass of an object at the Earth's surface.

b 90 kg; 342 N

c 9.8 N kg^{-1} = 9.8 kg m s^{-2} kg^{-1} = 9.8 m s^{-2}.

6 a

moment = $F \times d$

b 4000 N m

c $T \cos 50° \times 3$ m = 4000 N m, so

$$T = \frac{4000\,\text{N m}}{\cos 50° \times 3\,\text{m}} = 2074\,\text{N}$$

7 a 40 MPa

b 0.2 (or 20%)

c 200 MPa

8 a
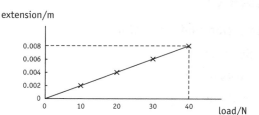

Spring constant = 5000 N m^{-1}.

b the load beyond which the spring becomes permanently strained

c 0.16 J

9 a See diagram on page 22; motive force is forward frictional force of road on tyre.

b reduced friction between tyre and road

c 5 m s^{-2}

d 100 m

10 $\Delta E_{p} = 1.96 \times 10^{9}$ J; $\Delta E_{k} = 8.75 \times 10^{8}$ J

11 a 3.3 m

b 22.3 m s^{-1}, vertically downwards

c 7.1 m s^{-1}, vertically downwards

d 3.1 s

12 a newton (N)

b 1 N = 1 kg m s^{-2}

13 a See graph on page 21.

b Elastic while straight line, plastic when curved (beyond elastic limit).

c Elastic: returns to original dimensions when stress removed.

Module B: Electrons and photons

1 **a**

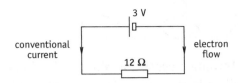

b 0.25 A; 0.25 C

c 3 V

d 0.75 J

2 **a** $V = W/Q$; p.d. = energy transferred per coulomb.

b $1 V = 1 J C^{-1}$

c $W = 1 J$; $Q = 1 C$

3 **a** in parallel

b 5 V

c 1250 Ω

d increase

e stay the same

4 **a** 1.52 Ω

b

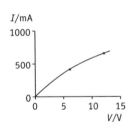

The I–V graph is not a straight line through the origin, so the copper alloy wire is non-ohmic.

5 **a** 5.0 V

b 10.0 V

c smaller current, so fewer lost volts

6 **a** 66.7 Ω

b 3.6 kWh

7 **a** 0.0144 N

b 3.4 A

c Measure force by balancing; knowing B and l, can find I.

8 **a** in free space, or in a vacuum

b 5.0×10^{-7} m (= 500 nm)

c visible

9 **a** single wavelength

b 4.14 eV

c the minimum energy needed to remove a conduction electron from the metal

d 1.94 eV

e Only the most weakly bound electrons have this KE. Most of the electrons were more tightly bound within the metal.

10 **a** *brief description of electron diffraction – see page 52*

b 2.2×10^{-14} m

Module C: Wave properties

1 Angle of incidence = angle of reflection.

Incident ray, reflected ray and normal to surface all lie in same plane.

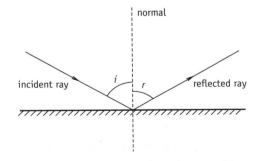

2 **a** 1.467

b 43.0°

c 30.1°

3 **a** 1.99×10^{8} m s^{-1}

b Angle of incidence > critical angle (41.5°).

c (i) 190 m

(ii) 9.5×10^{-7} s

d Signals are dispersed ('smeared') as they travel along, because some parts take longer than others.

e All rays travel same distance along fibre.

4 a 0.12 m

b increased

5 Metal plates vertical with slit, width a few cm, between them. Generator behind slit. Explore space beyond slit. Look for high and low intensity points.

6 a A bright fringe is formed where two waves, one from each slit, meet and interfere constructively. The path difference between them must be a whole number of wavelengths.

b 1.3 mm

7 a sound waves

b light waves and microwaves

8 a Node: amplitude of vibration is zero. Antinode: amplitude is maximum.

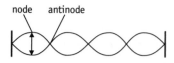

b Travelling waves run along string in both directions. Waves reflect at ends, and reflected waves meet and interfere with each other.

9 a Transverse, as displacement perpendicular to velocity.

b 1.2 m s^{-1}

c 2 cm

Index

A

acceleration 12–13
  definition 4–5
  signs 4
  units 4
accuracy 76–7
air columns, waves 73
ammeters 32
ampere (A) 44–5, 28, 30
  definition 44
  derived units 45
amplitude (A), waves 64
antinodes 72
area 5
attraction, magnetic 40

B

braking forces 23

C

calculating electrical power 38
centre of gravity 19
changing wavelength 63
charge (Q) 28
checking units 7
circuits 34
  symbols 82
coherent sources 69
combining electromotive forces 35
compression 20
connections
  parallel 28
  series 28
conservation of current 28
conventional current 28
corkscrew rule 41
coulomb (C) 28
couple 19
critical angle C 60
current (I) 28–9, 39
current-carrying conductor, forces 42
current-voltage characteristic graph
  32

D

data 78–9
  transmission 61
de Broglie equation 48

definitions
  acceleration 4
  resistance 30
  useful 80–1
deforming solids 20–1
density 19
derived units 13
diffraction 68–9
  electrons 52
  explanations 69
  ripples 69
displacement 2–3
  waves 64
displacement-time graphs 3
distance 2
doing work 16
drag 14

E

elastic deformation 21
elastic limit 20
electric circuits 34
electric current (I) 28–9
electrical circuit symbols 82
electrical power 38–9
  calculations 38
  units 38
electrical resistance 30
electromagnetic radiation
  production 47
  speed 47
electromagnetic spectrum 46–7
electromagnetic waves 62
electromagnets 40
  flat circular coils 41
  long straight wire 41
  solenoids 40
electromotive force (e.m.f.)
  34–6
  combining 35
  formulae 34
  meaning 34
  measurements 36
  units 34
electron diffraction 52
electrons 27–56
electronvolt (eV) 48
e.m.f. *see* electromotive force

energy
  electromagnetic 46
  photons 48
  wavelength 46
equations of motion 6–9
  derivation 8–9
  the four equations 6
equilibrium 19
errors 76–7
explaining diffraction 69
expressing errors 76
extension 20

F

field lines 40
field strength, gravitational 14
flat circular coils, electromagnets 41
Fleming's left-hand rule 42
flux
  density (B) 40
  lines 40
force on a current-carrying conductor
  42
forces
  acceleration 12–13
  motion 1–26
  unbalanced 12
  vehicles 22–3
  work 16–17
formulae 78–9
  Q, I and t relationship 29
free fall 14
frequency
  measurements 65
  photons 48
  waves 64, 66
fundamental units 13, 44

G

GPE *see* gravitational potential energy
gradient 3
gravitational field strength 14
gravitational potential energy (GPE) 17
gravity 14

H

hertz (Hz) 64
Hooke's law 20

I

interference 68–9
  fringes 70
  sound 68
internal resistance (r) 36–7
  measurements 36

J

joule (J) 16

K

kilowatt-hour (kWh), energy unit 39
kinetic energy (J) 16
Kirchhoff's first law 28
Kirchhoff's second law 35

L

laws of reflection 58
laws of refraction 58
longitudinal waves 62

M

magnetic attraction 40
magnetic fields 40–1
magnetic forces 42–3
  calculating 42
  current-carrying conductor 42
  explaining 42
  usage 42
mass 12–13, 14
  meaning 13
measurements
  electromotive forces 36
  frequency 65
  internal resistance (r) 36
  resistance 32
  wavelength of light 70
microwaves 73
  Young's slits experiment 71
moment of a force 18
momentum, photons 49
motion
  forces 1–26
  straight line 2
motive
  force 22
  power 22
multipath dispersion 61

N

newton (N), unit 13

newton-metres (Nm) 18
nodes 72
non-ohmic conductors 32
normal 58
north pole 40

O

ohmic conductor 32
Ohm's law 32–4
  non-ohmic conductors 32
  resistivity 33
  temperature dependence 32
ohms (Ω) 30
optic fibres
  advantages 61
  practical fibres 60
  total internal reflection 60
  transmitting data 61

P

parallel
  connections 28
  current attraction 44
  resistors 28
pascals (Pa) 21
path difference 68
p.d. *see* potential difference
pencil sharpener rule 41
period (T), waves 64
permanent magnets 40
phase difference 64
photoelectric effect 50–1
  explaining 50
  observing 50
photons 27–56
  energy 48
  momentum 49
  photoelectric effect 50
pivots 18
Planck constant 48
plastic deformation 21
polarisation 62
potential difference (p.d.) 34–6
  formulae 34
  units 34
potential dividers 36–7
  varying resistance 37
power (P) 38
  work 16–17
pressure 21
principle of superposition 72

progressive waves 66
projectile motion 15

R

radians (rads) 64
random errors 76
rays 63
reflection 58–9, 63
refraction 58–9, 63
refractive index (n) 59
repulsion, magnetic 40
resistance (R) 30–1
  measurement 32
resistivity, Ohm's law 33
resistors
  in parallel 31
  in series 30
right-hand grip rule 40
ripple diffraction 69
ripples, Young's slits experiment 71

S

scalar quantity 3
series
  connections 28
  resistors 28
SI units 39, 44
significant figures 77
signs, acceleration 4
sine waves 62
Snell's Law 58, 59
solenoid 40
solids, deformation 20–1
sound interference 68
sound waves 62
speed 2
  waves 66
speed of light (c) 47
spring constant (k) 20
standing waves 72–3
  conditions 72
  stretched string 72
stationary waves 72
  conditions 72
stopping distances 23
straight line motion 2
strain energy 20
stress-strain graphs 21
stretched wire 20
superposition, waves 72–3
systematic errors 76